CABBAGE
OR
CAULIFLOWER?

CABBAGE
—OR—
CAULIFLOWER?

A Garden Guide
for the Identification of
Vegetable and Herb Seedlings

JUDITH ELDRIDGE

David R. Godine · Publisher
BOSTON

First published in 1984 by
DAVID R. GODINE, PUBLISHER, INC.
306 Dartmouth Street, Boston, Mass. 02116

LIBRARY OF CONGRESS CATALOGING IN PUBLICATION DATA

Eldridge, Judith.
    Cabbage or cauliflower?
    1. Vegetables—Seedlings—Identification. 2. Herbs—
Seedlings—Identification. 3. Vegetables—Identification.
4. Herbs—Identification. 5. Seedlings—Identification.
I. Title.
SB320.9.E43 1984   635'.0431   83-48520
ISBN 0-87923-497-0 (pbk.)

PRINTED IN THE UNITED STATES OF AMERICA

This book is dedicated, of course, to William.
I would also like to thank Valerie.

Contents

Cabbage or Cauliflower?

In a way, my garden looks its best in the early spring. It's then that the newly turned earth most resembles the map I made late last summer, when I thought I saw clearly where things really ought to be planted. The garden in April looks quite like my plan, although the dirt is an even damp dark and my paper is still white. The edges on both are still neat in April. The rows seem straight. It will be late June before I discover that I persist in planting the beans too close for easy picking, my rows are not quite in line, and I didn't follow the map anyway.

This book is for spring gardening. It won't tell you everything about where or when or how to plant. Or what to buy. It is for folks who are well armed with confidence, enthusiasm, and numerous other books about gardening. This is an identification guide to seedlings; it describes and illustrates all of the common and some of the not-so-common vegetables and herbs grown in North American gardens, from the time they first appear above the soil until they acquire their fourth or fifth true leaves.

It's for those of us who lose that quickly soiled garden map almost immediately.

A book for those who think they are going to remember what they've planted where—and don't.

For those who write labels with felt-tip pens, and then water them.

For beginners who threaten their newly sprouted carrots with an exuberant fit of weeding.

For adventurers who experiment by sowing less common seeds and aren't sure whether or not they were successful.

And for every forgetful gardener who fervently hopes that those sprouts aren't ALL going to be zucchini.

This book is an identification guide for the expert as well as the novice.

Somehow those garden charts that many of us create ahead of time lack relevance when we are faced with the garden itself. An eight-foot row just doesn't look the same paced out in the winter living room as it does in the spring garden. The temptation to ad-lib at planting time is strong, and last-minute additions often slip by unrecorded. The calculating approach with clean paper and pencil is in opposition to the dirty-fingered, whimsical method of garden layout to which most of us succumb. Thus the identification problems of June are often initiated in May.

Even after many seasons of planning and planting and harvesting, these brief reunions in early spring with our new plants often leave us hesitant. The face is familiar but the first name escapes you. The conversation is awkward at first. Is that the cabbage? Or is it the cauliflower? Perhaps a relative you've never met before. Eventually, one way or another, as the plants mature, their identities are sorted out and all is forgiven and forgotten until the next spring. The purpose of this book is to solve those early spring confusions when they occur.

There are several general observations to be made when beginning the study of vegetable and herb seedlings. The first is that an infuriating number of plants that are wonderfully dissimilar on the dinner plate look distressingly alike when first seen growing in the garden. Secondly, one notes that the first one or two leaves to appear seem rather mismatched or even misshapen compared to the ones that follow. Lastly, one observes that intended plants and weeds often seem identical and inseparable. There are rather precise botanical explanations for these occurrences, the understanding of which will make work simpler and life more interesting for the curious backyard farmer.

In order for any explanation to succeed, a description of how plants are scientifically named is necessary. Botanists early on saw the need for a universal method of describing each type of plant. Arguments about the best way to accomplish this went on for centuries but the general scheme finally adopted is easily understood.

The system of botanical nomenclature initially arranges plants under broad headings according to basic similarities. These large groupings are then divided and subdivided in an orderly fashion, aligning into ever-smaller groups plants that hold many characteristics in common. The internationally recognized name for each specific plant included in this book is for our purposes usually reduced to the last few of these classifications, a general family name, a genus name, species name, and, if necessary, a botanical variety name. The concise descriptive phrase is latinized, resulting in confusion and mispronunciation for most of us non-biologists.

There is something to be gained in overcoming the urge to skip over or ignore the foreign-sounding words. These names are the first clue to the relationships between the different cultivated plants and likewise between the crops and the weeds. Seeing that two plants have been listed together under these final classifications leads one to expect that even as seedlings their physical similarities will be great. One might not previously have expected a young beet plant, *Beta vulgaris,* to look at all like a youthful Swiss chard, *Beta vulgaris* var. *cicla.* Determining the family grouping will often assist in understanding the shared growth habits and requirements of plants with which you may not already be familiar. Very similar plants, as identified by these scientific names, may cross-pollinate, and this information can sometimes be very useful. So, confront the Latin; greater rapport with what is growing around your ankles results.

Fluency with the internationally accepted nomenclature is a must for any in-depth reading or research on horticulture. There comes a point where the common names take a back seat to the botanical names, or are absent altogether. Note, however, that there is still disagreement over the most accurate name for some of the plants included here. In these instances two or more scientific names are listed.

The illustrations in this book are gathered together under the large and general headings of the family names. The plants in each family grouping do not necessarily share common ancestors, as the word "family" might imply, or even share a common place of origin, although this is often the case with our vegetables and herbs. Plants within a family do have similar flower structures and seeds; they also tend to share similar needs and aspirations, pests and preferences, and often trade diseases with each other.

Each of these families is broken down into smaller genus groups containing plants with more closely shared similarities. A family may contain one or even hundreds of these genus headings. Each distinct type of plant in a genus then has its own universally recognized identifying name, the species name, which separates it from its fellow genus members. Not too surprisingly, different species of seedlings that are of the same genus and family appear to be identical at first glance and are easily confused.

Additional nomenclature, the "botanical variety" name, is used to separate two species of the same genus that are in minor ways botanically different from each other, their structural differences not great enough to warrant separate species names. The family, genus, and species names will usually suffice when referring to any particular wild plant. The taxonomy of cultivated plants, though, is often more complicated due to the tinkering of mankind, and there is frequently need of a botanical variety name to specify the slight variations that have been developed. This variety name, also latinized, is set off from the genus and species names by the abbreviation "var." It is almost always possible to distinguish these botanical varieties from their related species at the very early seedling stage. The exceptions are indicated in the text pages opposite their illustrations.

Uncultivated plants are fairly homogeneous; that is, the offspring of a given wild plant don't vary a great deal from each other or from the parent because any extreme variations are not usually able to cope successfully with the harsher environment. The history of many of our cultivated plants began when some of these interesting, naturally occurring deviations of wild forage plants attracted the notice of early agronomists and were aided and propagated in a somewhat more congenial setting than they would have found in the tangle of competing vegetation. In most cases, as our food crops became even more civilized and were actively bred for their usefulness to mankind and ease of cultivation, their ability to survive on their own diminished further or vanished altogether.

The early haphazard and the later more deliberate methods of plant breeding have produced crops that are genetically different from their undomesticated forebears. More recently, plants that would not normally have given each other the time of day, so to speak, have been gathered together and hybridized to create exotic new greenery. Engineered mutations have also added to the confusion of naming and specifying relationships between cultivated plants. Plants differing quite radically from the forms from which they were developed receive their own species name, and frequently a botanical variety name is needed to separate those plants which are only slightly different from each other. Some of the many interesting books and papers that cover in detail the history and effects of human intervention and cultivation on the plant world are listed in the bibliography.

Further distinguishing names are needed for the various forms that each species of garden plant may take under cultivation. There are, for instance, dozens of shapes, colors, and flavors of garden radishes. This is where those clever, colorful names invented by the seed companies and researchers come into use. For example, "Crimson Giant" radishes. Officially, this is called the "cultivated variety" name or "cultivar" name for that very specific form of the plant. The cultivar name for a plant is not necessarily recognized universally. Sadly, it is impossible for us as mere garden-weeders to differentiate between the various cultivars of a plant in these early phases unless the variety differences involve coloring or shape of leaf. For instance, we can easily separate a red "Rhubarb" chard from a green Swiss chard, or an "Oak Leaf" lettuce from a "Black-Seeded Simpson," but we cannot tell if a seedling carrot is going to be a "Red Cored Chantenay" or an "Oxheart" just by viewing the first leaves. In this case it is solely the shape of the mature root that makes the difference between these cultivars.

The botanical family *Cruciferae* provides a classic example of the diversification of cultivated plants. The cabbage-related plants of the genus *Brassica* of this family are among the most easily confused seedlings in the spring garden or coldframe. *Brassica oleracea*, the original field cabbage, of the lands of the Mediterranean and Asia Minor, was smaller, loose-leafed, and quite bitter-tasting. By selecting and saving for seed only the best plants, the early agriculturalists of these regions developed the kales, which have tender, more numerous leaves and a better flavor. Collards were the next distinct vegetable crop to appear through this continual slow search-and-selection process, the leaves larger and thicker than those of the kales.

Eventually the hard-heading forms of cabbage were developed, still several thousands of years ago. Changes occurred not only in the leaves; occasional mutations in the stems and flower clusters were also sought out and exaggerated. Thus arrived the kohlrabi with its thick and edible stem, broccoli and cauliflower with edible flower buds, and finally, around 1590, the Brussels sprout with its miniature "heads" forming in the leaf axils. Still other cabbage contortions were once in use but are now out of favor. The search for larger, better-flavored, and disease-resistant varieties continues. Although the harvested products are different, the structure of the flowers, fruits, and seeds of this whole group of cabbage-derived plants has remained consistent. For that reason these vegetables all continue to share the same genus and species name with the original bitter field cabbages, and only a botanical variety name distinguishes them from each other.

To further illustrate the naming of plants, the entire botanical name for a particular Brussels sprout plant could be written: *Brassica oleracea* var. *gemmifera* "Jade Cross." "*Brassica oleracea*," the genus and species names, are the classical Latin words for cabbage and potherb. "*Gemmifera*" translated means "bearing buds" and is the botanical variety name, used here only for Brussels sprout plants. "Jade Cross" is the cultivar name for that specific form of Brussels sprout.

Family resemblance begins early. The very first green shoots that break through to the garden surface can tell a great deal about the forthcoming plant and what family category it might belong to. This knowledge is often a clue to its future growth habits and requirements. Those distinctive heart-shaped leaflets, for instance, indicate that what has appeared above ground belongs to the family *Cruciferae*. Two tiny, sparkly first leaves mean that either you have invading carpetweed or your New Zealand spinach is up. Or both. They are rather closely related also.

Curiously, these initial green unfurlings, the first simple-shaped leaves, often bear no resemblance to the leaves that unfold thereafter. The reason for this lies in the anatomy of the seed itself. All of the vegetable and herb seeds discussed herein, be they boring little brown spheres or more colorful and elaborate affairs, have three basic parts: the embryo, food storage tissues, and seed coverings. The embryo, which is the new plant itself, consists of an axis with the beginnings of a root at one end and a green shoot at the other. One or more "seed leaves" or cotyledons are attached. These cotyledons usually contain enough stored food to sustain the incipient plant until its roots and leaves can take over for themselves. When they are visible, they look rather like simple green leaves, although the "real" leaves of the plant usually turn out to be quite different in shape.

A few of the plants included in this book leave their cotyledons under the surface of the soil when they emerge. Peas and corn are good examples. For those who are fond of official-sounding words, this is called hypogeous germination. Those plants that display their cotyledons above the soil are demonstrating epigeous germination. Most vegetable and herb plants employ this latter method and sport them as regular leaves, at least for a while. They shrivel and drop off when they are no longer needed.

Most of the plants illustrated here are dicotyledonous, meaning that they have two such "seed leaves." The rest are monocotyledons and have only one—onions, for example. Another feature which distinguishes monocots from dicots, whether or not the cotyledons are visible, is the fact that the veins of the monocots' leaves parallel each other, rather than branch outward from central veins as do those of the dicots. Again, plants belonging to the same botanical family will have very similar cotyledons.

Vegetable and herb plants are not only easily confused with each other, but also with the inevitable weed plants. These encroaching look-alikes may belong to the same botanical family as the intended crop. Weeds, of course, are defined as anything that isn't deliberately placed to grow within the prescribed limits of the garden. Thousands of years ago our garden plants were themselves weedlike in the sense that they took care of themselves and were not intentionally cultivated. Some still bear a remarkable resemblance to their ancestors. The carrot, for example, is only slightly removed from one of its progenitors, the Queen Anne's lace.

As a rule, weeds are not clever enough to align themselves in rows and thus disguise themselves as crops. However, many gardeners cannot plant in straight lines, either. Some, lacking confidence, adopt a "wait-and-see" attitude before weeding or thinning the uncertain parts of their gardens. Apparently they hope that their intended plants will be easier to find when the weeds are taller. This waiting period is quite unnecessary, though, as positive identification can be made by studying the structure of the young plants carefully.

If we define weeds as unintended additions to the garden, the definition then also includes those plants sprung from seeds left lying about by last year's careless vegetables and herbs. Stray seeds hardy enough to survive the winter and the rototiller may appear voluntarily the following spring. Sometimes they have done you a favor by replanting themselves, sometimes not, depending on who their parents were and where they've arranged themselves.

The plants that concern us here use one or another of two basic methods for forming seeds. By one method, the seed-producing flowers are self-pollinating; they will only accept their own fertilizing pollen. A large number of our garden plants, though, can readily accept the pollen brought by wind or insects from another plant. This outside source must, however, be of the same species, although it can be of a different botanical variety or cultivar of that species. Problems sometimes arise when different varieties of the same species cross-pollinate in this way. If hardy seeds from this "intermarriage" of similar but different plants have escaped in the garden, or come back in with the compost, they may produce plants next year that are a confusion of the characteristics of both parent plants and not worth the watering, weeding, and debugging. Seedlings of these mixtures will look familiar at a glance but under scrutiny will appear slightly different than either parent.

The squash family is forever putting me in this awkward position. No matter how attentive I am, every year a handful of seeds from last year's leftover vegetables reappear as seedlings in spring, looking robust and promising an early and bountiful cucurbit vine. It seems such a shame to murder them. In my kind-heartedness, I have on occasion carefully nurtured vines that proceeded to produce some interesting but not particularly flavorful gourd-like beasts. Nowadays I'm less nice to seedlings that I don't recognize.

An example might help to explain this trying situation. Zucchini and yellow crookneck squash are very different varieties of the same family, genus, and species, *Cucurbita pepo.* They tend to flower at the same time and are not particular about whether they pollinate themselves or trade pollen with others of their species. The results do not appear until the seeds of these unions are sown in the next planting season, because this year's edible portion, the ripened ovary surrounding the seeds, will not be affected by any pollen change. On the other hand, a zucchini will not have anything to do with, for instance, a watermelon. The two appear quite similar in early spring, and are of the same family group, but they are of different genera and species, which means that they are not enough alike to mingle in so intimate a manner.

Many groups of plants commonly found in our gardens, which would intermingle if given the opportunity, are normally harvested before they flower and set seed. Additional information about various plants that tend to crossbreed and then resow themselves is given on the pages facing their illustrations.

Many of the seeds we buy today are pre-arranged or even forced cross-pollinations of two inbred varieties of plants, each of which has some favorable characteristics but is not all that great by itself. Plant breeders have found that by playing matchmaker to varieties that may not have combined on their own they can produce seeds that will in turn produce plants far superior to either parent, combining the desirable qualities of both and getting the benefit of that mysterious phenomenon known as heterosis, or hybrid vigor.

Some of these hybrid seeds are produced by planting the desired varieties near each other and letting wind or insects carry on the pollen-switching. Other crossings, especially those involving plants that would normally pollinate their own flowers, require more drastic measures to insure that only the pollen of one special variety will fertilize the variety that is to produce the seed. The extra time and effort involved in these crosses causes these hybrids to be more expensive than the non-hybrid varieties of the same plant. Seed packages and catalogues clearly state which of their offerings are hybrids, and frequently the notation "FI hybrid" is added, meaning that the plant produced from this seed is the first generation to come from that crossbreeding.

The intricate workings of chromosomes and dominant and recessive genes give us the same sorts of problems with the offspring of intentional hybrids that we have with the offspring of unintended crosspollinations. The SEEDS of these hybrids do not necessarily produce plants that are as nifty as the hybrid parent. Less favorable traits that were recessive with the first or FI generation may reappear with the next generation.

For example, you may have had simply magnificent results with your "Early Girl Hybrid" tomatoes last year, these "Early Girl" seeds being the result of a carefully managed cross-pollination of two different varieties of tomato. Volunteer tomato sprouts that come up the following season from mislaid fruit of this hybrid may not perform quite as well as last year's tomato plants. This is because they have reverted to the less than ideal characteristics of the original varieties that made the hybrid. Seeds from any plant that was itself a hybrid will never give consistent results.

If you had instead planted one of the standard or "old-fashioned," non-hybrid varieties, such as "Marglobe," its potentially overwintering offspring would resemble the original. This is because tomato flowers if left to themselves are self-pollinating, and the non-hybrid offspring will consistently resemble the parent plant. If they have chosen a convenient spot for themselves, or can be moved to one, you ought to consider letting them grow.

If you are interested in cultivating some of your volunteer vegetable and herb seedlings, it is a good idea to think back on what related varieties you grew last year and whether their tendencies are to trade pollen. Attention to the species and variety names is important. Books devoted to seed production and seed saving offer further details on the vagaries of the different species.

It is sometimes useful to know just when one might expect to see the seedlings emerge. A general expected arrival date gives some idea of whether you should be concerned about the seedlings' health, the age of the seed planted, dirty tricks the weather may have played, and whether or not it's time to give up hope and replant. Sometimes this emergence date is even useful as an identifying feature. For instance, if you are elated that your celery seems to have appeared in three days, you should be aware that it is probably not really the celery. Even under the best of conditions celery seems to take forever to come up.

In order for the expected dates of emergence to be effective, the planting date must be legibly, indelibly, and retrievably recorded somewhere. This is where I usually come apart, but in this book I have included, for those more organized than myself, the anticipated time of seedling appearance along with the illustrations.

Many gardening-book and seed-package writers erroneously use the phrase "germination date" to mean the emergence date, the time when the seedling is first visible above the soil. The actual process called germination begins at the moment a viable seed begins to absorb water through its seed coats. From this point of water acceptance all sorts of intricately interrelated metabolic changes occur. Enzymes bestir themselves, changing stored food into usable solutions; hormones reactivate; cells enlarge and/or multiply; respiration increases . . . all before a root or shoot breaks through the now water-softened seed coat, and long before the new plant reaches the notice of its fond but anxious depositor.

The success of any new planting depends first on the age of the seeds and the conditions under which they were stored. Many people who are new to this business of food-growing are under the impression that seed must be purchased fresh each year. Perhaps this is how some people manage to have empty seed packages to impale on sticks at the end of each row—they simply discard the unused seed. With the exception of a few species whose seed remains viable for only one or two years, these being onions, parsnips, corn, salsify, soybean, peanuts, parsley, and a few other herbs, dry seed stored in a cool, dry location should last for at least three years.

Seedsmen recognize the longevity of seeds and collect the unsold packets from the shelves in late summer. Someone then sorts them all out, and if the seed continues to pass strict germination tests, it may be repackaged and offered for sale the following season.

After you have contemplated for a moment the age and potential viability of the seeds, it is time to commit them to the planting medium—in protected flats or pots or directly in the field. The next considerations are water, warmth, and air: three essentials for seed germination. Soil temperature is more significant at this stage than air temperature; in general, these vegetable and herb plants prefer a range of 75–90 degrees F. (24–32 degrees C.) for optimal germination. The accessibility of water, as mentioned earlier, is absolutely necessary to begin this process, but if the seeds are continually flooded with water the equally necessary supply of gases in the soil will be cut off. Oxygen deprivation may also occur if seed is planted too deeply or in a very compacted growing medium. These are the major variables that can be manipulated by gardeners, and they will influence not only the speed with which the plant appears, but whether or not it appears at all. The requirements for particularly idiosyncratic species are noted on the pages opposite their illustrations.

The day will arrive when the seedlings make their presence known. The time has come to sit back and try to recollect whether these seedlings are what you were expecting to see. The same environmental factors that influenced their germination will, of course, affect their continued good growth. Extremes in growing conditions can alter the way a seedling looks, the ease with which it can be recognized if its identity is in doubt, and thus the usefulness of this identification guide.

One does not often see in seedlings contortions so dramatic that they affect the plant's ability to be identified, but the list of potential problems is intimidating. To state the obvious, plants tend toward an unnatural horizontal position to express thirst or disease. Another obvious contortion, one with which windowsill farmers are all too familiar, is the stretched, leaning posture of a seedling that has been given insufficient or unbalanced light. The stem distance between the leaves will be elongated, the leaves pale and undersized.

There are also invisible soil conditions that can affect seedling appearance. Excesses or deficiencies of the essential nutrients—nitrogen, phosphorus, and potassium—will change the growth habits and the color of the foliage. Yellowing or browning of the leaves, spotting, darkening of the veins, curling of the leaf edges, stunting of growth in general—all are common symptoms of trouble, varying with the species of plant. Soil acidity may also have some effect on seedling appearance, as will extremes in air temperature and humidity, over-fertilization, virus and fungus attacks, and insect infestations.

Commercially prepared potting soils, for those vegetables and herbs that are started early under lights, in coldframes, or on bright windowsills, can be counted on to be well-balanced. In established gardens, chances are that any drastic imbalances will have been noticed in previous seasons with disappointing harvests, and subsequently corrected. Professional soil analysis can take the mystery out of soil problems.

The illustrations which follow are of seedlings grown under almost ideal conditions. An attempt was made to select as models the most average-looking plants rather than the rare Super Plants. All plants and seeds are drawn life-size (except where noted) but keep in mind that "normal" covers a wide range of size with seedlings. Plants may be somewhat larger or smaller than those depicted and still be perfectly healthy.

In general, the shapes of the first leaves of all of the vegetable and herb seedlings included here are just a hint of what is to follow. The later leaves of most plants will look similar but larger and more complex than the seedling leaves depicted. However, a few plants, okra and zucchini for example, change radically as they grow older. I have included descriptions of the appearance of mature plants in cases where their identification might be a problem. The cotyledons, if they were once visible, may have withered and fallen off, usually by the time the fifth or sixth leaves have appeared. The first, or lowest, leaves may also drop from a healthy plant as it matures, making it difficult to recognize from its illustration as a seedling.

A note is necessary here about that most disturbing family, *Cucurbitaceae.* This family includes all of the many species and varieties of squashes and pumpkins, as well as all melons, cantaloupes, cucumbers, and watermelons. It would be impractical to include all of the vast and increasing number of types now offered for sale. Many cucurbit vegetables that produce very different results at harvest look exactly the same as seedlings. The only logical way to handle identification problems of these closely related plants was to describe the most popular varieties and illustrate examples of plants that show obvious differences as seedlings. Some identity confusions can then be resolved through the process of elimination.

I hope that this book will offer you some assistance at this brief and worrisome stage in the life of your seedlings. No more should you have to suffer self-recrimination at haphazard planning or embarrassment at forgetfulness; no more have to accidentally weed away diminutive seedlings or otherwise mistreat plants due to confused identities. No matter what you'd planned to sow there, should have grown there, or thought was sown there; these pages will tell you what IS growing there.

CABBAGE
OR
CAULIFLOWER?

Sweet Corn   Popcorn

Gramineae   *Gramineae*
  Zea mays var. *rugosa*   *Zea mays* var. *everta*

Corn is known by its Native American name of "maize" in nearly every country but the United States. "Corn" is really an old Anglo-Saxon word which can denote a grain of any kind.

This particular grain probably began its involvement with mankind at least four thousand years ago in the Andes region of Bolivia and Peru. Corn underwent several major changes as it made its way north with wandering peoples to Central America and then across the North American continent. Recently, a small stand of a primitive perennial corn was discovered in southeastern Mexico.

Popcorn has been under cultivation for as long as the earliest "dent" and "flint" varieties that are used for flour and stock feed. The American natives served popcorn to the early European colonists at their Thanksgiving feast in 1630.

Our familiar sweet corn has been popular for little more than a hundred years. It no doubt occurred through the years as an occasional chance mutation, but was apparently not appreciated by the early cultivators, who did not bother to preserve it as a separate variety. It was only after the Civil War that several kinds of sweet corn began to be offered for sale.

Corn is quite obviously a member of the same botanical family that includes the grasses, and at the seedling stage it resembles some of these obnoxious weeds. Corn plants grow rapidly upward from a single thick stock, whereas the weed grasses concentrate on spreading horizontally, quickly sending up lateral shoots from the base or on runners. The slightly fuzzy stems on most weed grasses also give them away; corn is quite smooth and shiny in comparison. Different cultivars of corn may have reddish or purplish veining but all flavors and varieties—flint, dent, sweet, ornamental, and popcorn—are virtually identical in form when seen as seedlings.

All varieties of corn, with their fine windblown pollen, will readily cross-breed, affecting the flavor of this season's crop. The best method is to plant only a single variety or to stagger the time of planting so that the different types are not tasseling simultaneously.

Corn prefers a thoroughly warm soil and seedlings usually appear within seven days of the time of planting.

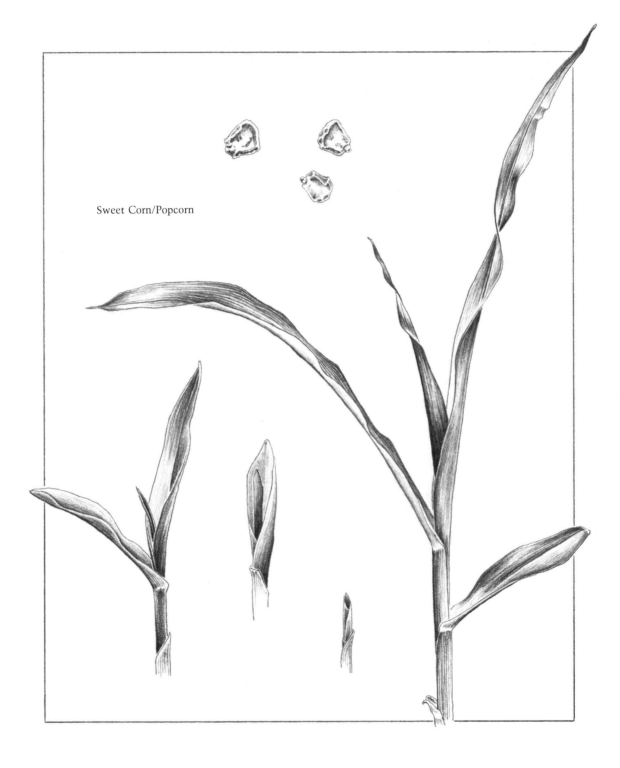

Sweet Corn/Popcorn

Asparagus

Liliaceae
  Asparagus officinalis

Mankind has been using asparagus for many thousands of years, first gathering the plant strictly for medicinal purposes. The original homeland of the asparagus has long been forgotten, as the plant escaped from gardens and naturalized itself over much of the Mediterranean lands and Asia before written language could record its passing. One of the most widely accepted theories as to its ancestral habitat places it in the eastern Mediterranean and Asia Minor.

Asparagus is unusual among our vegetables in that the pollen-bearing and seed-producing flowers are borne on separate plants. This is fairly irrelevant to the asparagus-eater, except that if said asparagus-eater is also the asparagus-grower, he or she may have noticed that some plants produce relatively fewer but fatter shoots than other plants. The thick-shooted plant is the female and will have red berries in the autumn; the male root is more prolific but produces thinner shoots.

The most popular method for starting an asparagus bed is to purchase year-old root stock, but many gardeners prefer to start some or all of their plants from seed. One advantage to sowing seed is that there are usually more varieties from which to choose; disease-resistant and newer forms are not always readily available as root stock. Another advantage is that one can start a number of plants and then single out only the sturdiest, most productive plants for the permanent bed.

Plants grown from seed are easy to overlook in the garden at first, and in its early stages asparagus can be mistaken for the fragile fronds of the field horsetail weed, *Equisetum arvense.* They are so thin that it is hard to believe they will amount to anything in the foreseeable future. Plants started from purchased year-old root stock are more easily recognizable. Shown on the right in the illustration are harvestable shoots.

Asparagus seed will germinate in about two weeks at a soil temperature of about 77 degrees F. and can be assisted by pre-soaking before sowing. The seeds are hardy and adventuresome and will sometimes turn up in unexpected places.

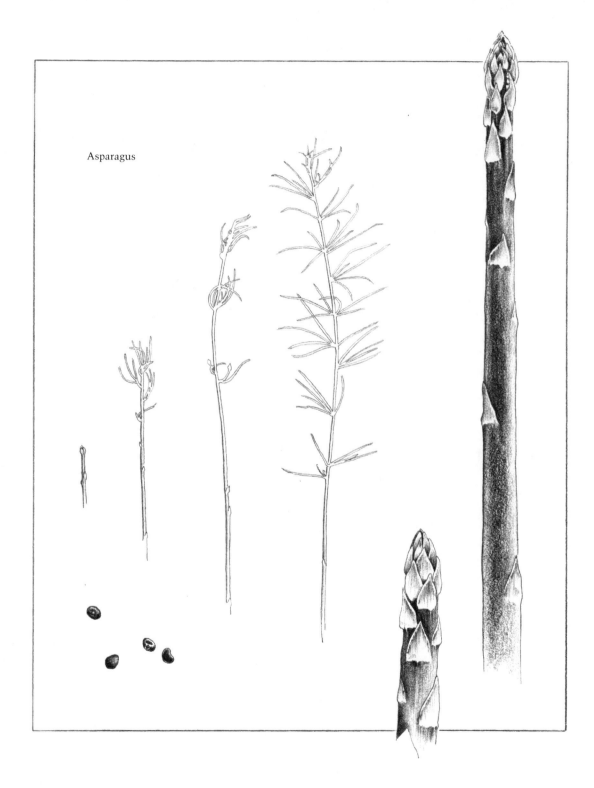

Asparagus

Onion   Leek   Shallot   Garlic

Amaryllidaceae *Amaryllidaceae* *Amaryllidaceae*   *Amaryllidaceae*
  Allium cepa   *Allium porrum*   *Allium ascalonicum*   *Allium sativum*

The plants of the genus *Allium* listed above were tamed and cultivated so long ago and have undergone such drastic changes through time that it is difficult to determine from what plants they may have evolved. Various guesses place the onion's origins in Asia Minor, the countries of the Mediterranean, or India. The ancient cultures of Greece, Rome, Egypt, and the Orient depended upon the *Alliums* for food and medicine, and attributed a magical and ritual significance to them. Onions are now grown in virtually every country and climate.

Common bulb onions can be grown either from seed or from dormant bulbs known as "sets." Onions grown from the dry sets send up robust, hollow, round shoots. The young plants grown from seed are so small and thread-like that they at first find tough competition from weeds and weeders who approach the task with excessive gusto. The seedlings look rather like thin grass weeds at a glance.

Scallion and leek seedlings look at first identical to the wire-thin seedlings of common onions. Scallions, also called bunching onions or green onions, are a form of *A. cepa* which do not form large bulbs. Maturing leeks are identifiable by their flat, ribbon-like leaves.

Garlic, too, has flat leaves. These appear as sturdy, curving stems with a neat crease in the center. Garlic is almost never grown from seed; indeed it seldom forms seed, but is propagated by breaking up a mature bulb into separate cloves and replanting.

The shallot is another "multiplier bulb." Rather than forming a single bulb, shallots form a cluster of triangular-shaped cloves. Like garlic, shallots are grown not from seed but from "sets." The first appearing foliage looks much like the large hollow leaves of an onion grown from sets, but shallots tend to send up a larger number of smaller shoots.

Onion and scallion seeds are not long-lived and won't remain viable for more than two years. Leek seeds may last three years. Plants started from sets or cloves should appear in about seven days; plants grown from seed will of course take a bit longer, appearing in about ten days.

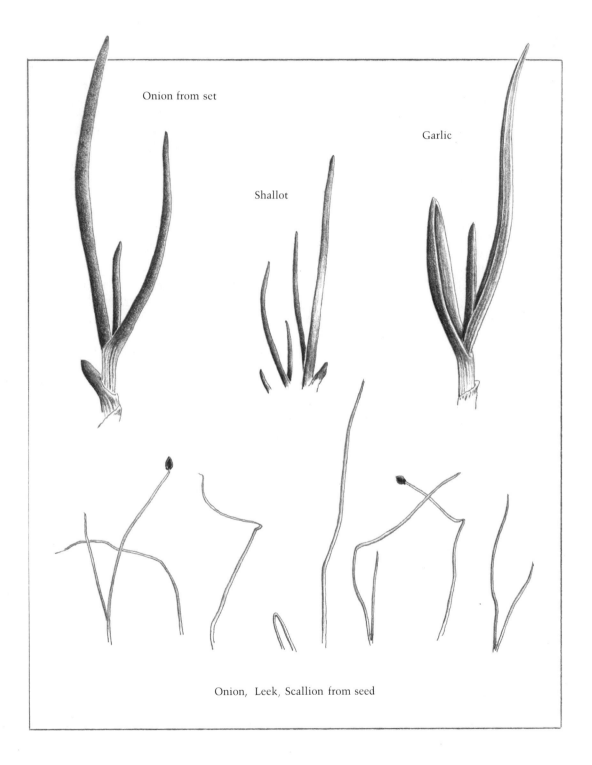

Onion from set

Shallot

Garlic

Onion, Leek, Scallion from seed

Chives

Amaryllidaceae
  Allium schoenoprasum

Chives were probably first cultivated in the Orient at least five thousand years ago. They are unusual among the herbs in that they were apparently not used extensively for any specific medicinal purposes but were savored primarily for their flavor and for their decorative effect. Chives were commonly grown in European gardens by the sixteenth century.

Chives are quite obviously related to the onion, scallion, shallot, and leek described earlier, and as young seedlings they are exactly alike. They sprout readily from seed and first appear as a single delicate thread, occasionally carrying the spent angular black seed case aloft. Except for their strong onion smell when bruised or cut, young chives could be mistaken for annoying grass weeds. On closer examination, it is clear that chives, like their relations, have hollow, round leaves whereas the grass weeds have flat, center-creased leaves. Chives are perennial, and an established clump will announce its presence in spring by sending up somewhat thicker, hollow shoots.

Seedlings should appear about ten days after sowing. As with onion and other onion-related species, the seed does not remain viable for long. Seed more than two years old may germinate poorly or not at all.

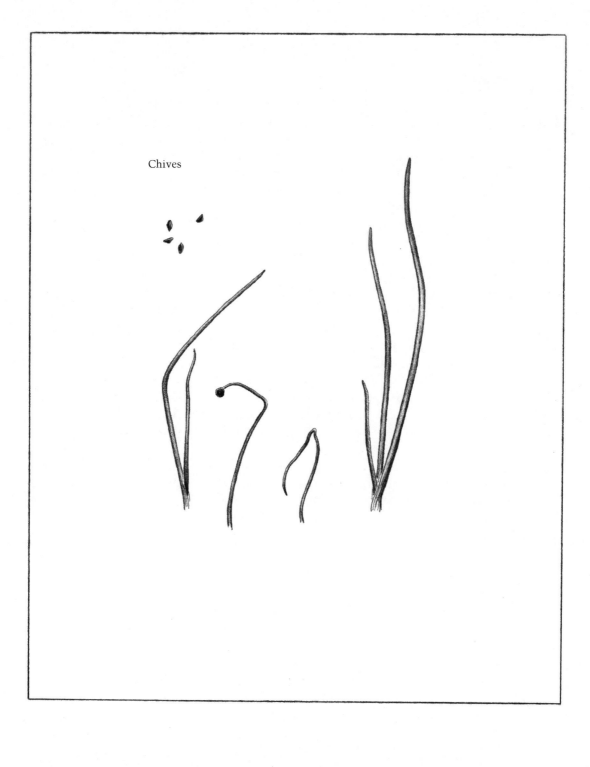

Chives

Swiss Chard

Chenopodiaceae
  Beta vulgaris var. *cicla*

Swiss chard, also called "leaf beet," was the earliest of this genus to be used extensively as food. Forerunner to the garden beet, it is essentially a beet without an enlarged root. Swiss chard came into use as a distinct vegetable crop in the vicinity of the North Sea and the Mediterranean, where the leaves had been gathered from the wild plants and used as a potherb since ancient times. The Greeks and Romans were the first to record the use of yellow and green varieties improved from the foraged plants, and Aristotle in the fourth century mentioned a red variety which is now commonly known as rhubarb chard.

The first appearing shoots of Swiss chards and root beets are very much alike; differences become more apparent with the first true leaves. Both plants send up their leaves in a similar cluster from the base, but the beet leaves are flatter and tougher than the very tender and puckered leaves of the chard. Chard has an only faintly reddish or pinkish stem whereas most varieties of beet will have a very bright red stem as well as a dark red-green tone to the leaves. Much later on, of course, the beet will show a thickening root.

   The early resemblance of the red variety of chard to the red garden beet is even more pronounced due to the greater similarity in coloring. There is also an unusual yellow beet which will more closely resemble the green chards. But again, the chard's fragile, curling leaves will soon give its identity away.

Like the beet, chard is quite hardy and seed can be sown directly in the garden in early spring, even before the last expected frost. It will germinate slowly if the soil is still very cold, but it usually appears in eight days.

Swiss Chard

Beet

Chenopodiaceae
  Beta vulgaris

The root beet, or common garden beet, is a comparatively recent development of Swiss chard and was probably first cultivated by the Romans in the second or third century A.D. *Beta vulgaris* var. *maritima,* the wild progenitor of the chard, root beet, sugar beet, and mangel, has a much smaller root and a bitter taste. Sugar beets and feedstock beets, not grown in ordinary backyards, still contain traces of these ill-tasting substances, as there is little reason to develop these crops for flavor. Garden beets were of relatively little importance as a food crop until after the eighteenth century, when improvements in form and flavor were developed in France and Germany, thus increasing their popularity.

Because the beet's history is so closely entwined with that of the Swiss chard, it is not surprising that their resemblance as seedlings is strong. The chard most commonly grown in North America, however, has very bright green leaves and only a slightly pinkish-red stem, whereas the beet is very dark red in stem and vein and has dark red-green leaves. There is, though, a red or rhubarb chard which is more easily confused with the beets, and a yellow variety of beet which more closely resembles the common Swiss chard. The best clue for sorting out these less common seedlings is that the chard leaves are always very tender and puckered and the beet leaves are always quite flat and have a tough, smooth feel.

Beets are a bit perverse in that what is commonly thought of as the beet seed is really a seed ball containing two to six small brown seeds. This means that for every seed that you think you've planted, in ten or twelve days two or more seedlings may emerge.

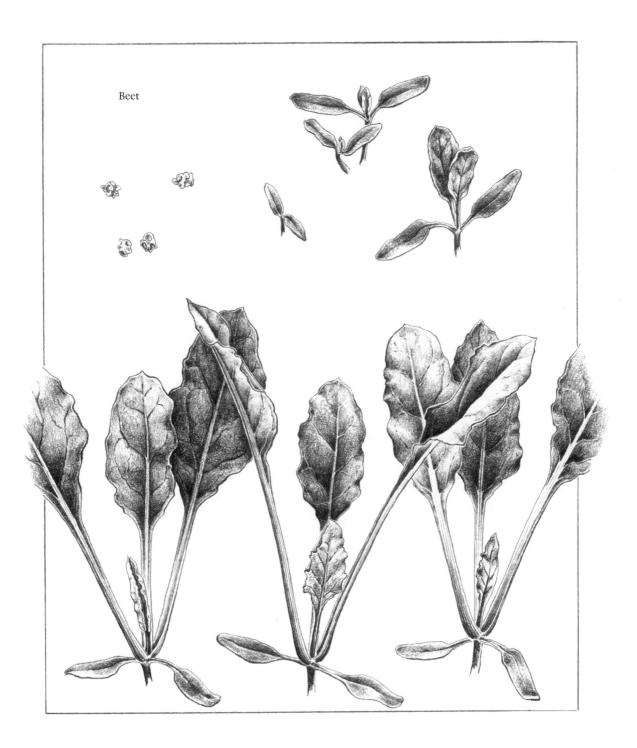

Beet

Spinach

Chenopodiaceae
  Spinacea oleracea

Spinach originally grew only in the area of what is now known as Iran, where it has been cultivated for several thousands of years. It was unknown outside its native land until about the first century A.D., when it seems to have made its way with peoples traveling from India to China. To the west, the merits of spinach were discovered somewhat later by other civilizations of the Mediterranean region and North Africa. It was grown in Spain by the twelfth century and is now almost universally cultivated.

The first appearance of these seedlings is made by a pair of long, narrow cotyledons that curl about quite gracefully and unmistakably. The next pair of dark green leaves to show are round and crumpled. This undulating leaf is reminiscent of the tender seedling of Swiss chard, a near relative. Swiss chard leaves tend to grow in a more upright fashion as compared to the low clumping of the short-stemmed leaves of spinach. The crinkled leaf varieties of spinach are by far the most popular in North American gardens, but smooth-leaf varieties are not un-usual.

Spinach is a cool-weather crop. Its seeds can even be planted in fall to germinate in spring. They will appear in about eight days when sown in spring. Spinach seed does not usually remain viable for more than three years.

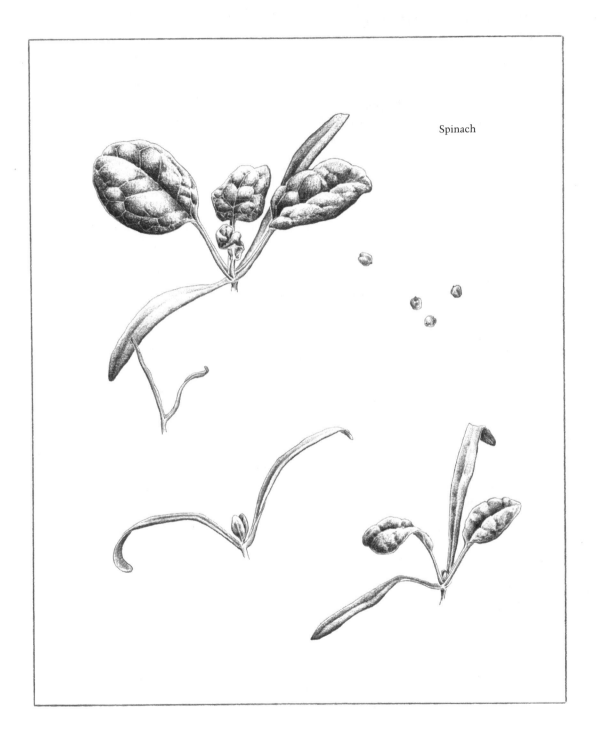

Spinach

New Zealand Spinach   Malabar Spinach

Aizoaceae
  Tetragonia expansa

Basellaceae
  Basella rubra or *B. alba*

Both of these foreign potherbs are slowly gaining acceptance in our gardens. These vegetables produce leaves abundantly in warm weather when our common spinach rapidly passes its prime. Varieties of New Zealand spinach have been cultivated in Australia, New Zealand, and Japan for quite some time. Malabar spinach, sometimes known as Basella or Ceylon spinach, originated in India and the tropics of the Far East, where it has been an important potherb for centuries.

New Zealand spinach is a low-growing, sprawling herb, an obvious relative to our familiar carpetweeds. This seedling first appears as two small, curling leaflets. The next leaves to emerge are triangular and curled backwards. New shoots soon appear in leaf axils, and the seedling rapidly spreads in an ever-growing circle. The surface of the leaves of most varieties of New Zealand spinach look as if they are encrusted with silver grains.

Malabar spinach looks quite different, although it too is a tender, crawling plant. It is apt to climb, given the opportunity. Its cotyledons are broad, like those of the squash family, but they are smooth and shiny, as are the large oval leaves to follow. Two species of this viney plant are grown. *Basella rubra* has a red tinge on leaf and stem but it is structurally identical to the white Malabar spinach, or *Basella alba*.

New Zealand spinach should be planted directly where it is to grow. It is quite hardy in cool weather, and seedlings should appear in eight days. New Zealand spinach will resow itself if allowed to flower. Malabar spinach will take a little longer to emerge; expect seedlings in about ten days.

New Zealand Spinach

Malabar Spinach

Kale

Cruciferae
  Brassica oleracea var. *acephala*

Kale is a primitive form of cabbage that has been cultivated for many thousands of years. The bitterness it sometimes acquires when harvested in hot weather is reminiscent of the taste of its earliest forebears, the wild field cabbages of the Mediterranean region. The kinds we grow today are little changed from those first kales. Botanically speaking, kale is a collard with frilled leaves, although its flavor is somewhat different. Thus the species and variety name given for both of these vegetables is the same. Kale, also known as borecole, was recorded as a food crop in North America in 1669 but since it was popular in Europe, it is generally believed to have arrived in this country somewhat earlier—probably with the first explorers and European colonists.

Kale's first seed leaves, the heart-shaped cotyledons, are identical to those of the other field cabbage descendants of this genus and species. Kale also has the same curious blue-green color of its relatives. The most common varieties grown in this country have finely-cut leaves, and this characteristic makes it easy to identify kale seedlings early on. Mustard at first glance will also appear to have frothy-edged leaves, but on closer examination one notes that the surface of the greener mustard leaves is bristly and dimpled, whereas the kale leaves are smooth but curled.

Less popular varieties of kale have red, yellow, or flat leaves. Those types with uncurled leaves will more closely resemble the cabbages and especially the collards which, like kale, have a more elongated stem than the heading cabbage.

Kale is very hardy, and will survive even a northern winter with little or no protection, flowering and setting seed during its second spring. When sown directly in the field in spring, kale should appear in about ten days.

Kale

Cabbage   Collards

Cruciferae
  Brassica oleracea var. *capitata*

Cruciferae
  Brassica oleracea var. *acephala*

Although both are descended from common ancestors in the Mediterranean region, the history of cabbage and collards diverges somewhat. Collards are a very primitive form of cabbage similar to kale and were well known to the most ancient civilizations of the Mediterranean, Asia Minor, and Europe, having been taken there by the Romans or possibly the Celts.

Heading cabbage, on the other hand, was unknown to the early Greeks and Romans. This botanical variety was developed in the cooler parts of Europe, probably by Celtic or Nordic peoples in the thirteenth century. Savoy, or crinkled leaf varieties, were grown in England by the 1500s.

The green varieties of cabbage and collards are identical as seedlings. When they are several months old, however, the compact heading form of the cabbage differs from the looser appearance of the collards. At this more mature stage, when a dozen or so leaves have unfolded, cabbage leaves begin to curl in on themselves, the stem remaining quite short as compared to that of the collard. Collards are more often cultivated in the hotter climate of the southern states, whereas cabbages do better in the north.

The cotyledons of the members of the family *Cruciferae* included in this book are all virtually identical. In addition, the leaves of the cabbage-related vegetables, which include Brussels sprouts, kohlrabi, cauliflower, kale, and broccoli, have a similar blue-green color and slightly waxy, smooth look to them.

It is the Brussels sprouts that most closely resemble the cabbage/collard seedlings, especially when only one or two leaves are visible. The leaves of these varieties all have an initial roundness and slightly scalloped edges. After the third or fourth leaves have appeared, one can see that the cabbage/collard leaves are thicker, less uniform in shape, and have a more prominent central vein than those of the Brussels sprouts. Another distinguishing feature of the Brussels sprout is that its leaves tend to be cupped backwards rather than flat.

Kohlrabi and cauliflower leaves are flat like the leaves of the seedling cabbages and collards, but they are longer and less rounded, more spear-shaped.

Like most members of this tribe, it will take about ten days for the seedling to emerge after sowing.

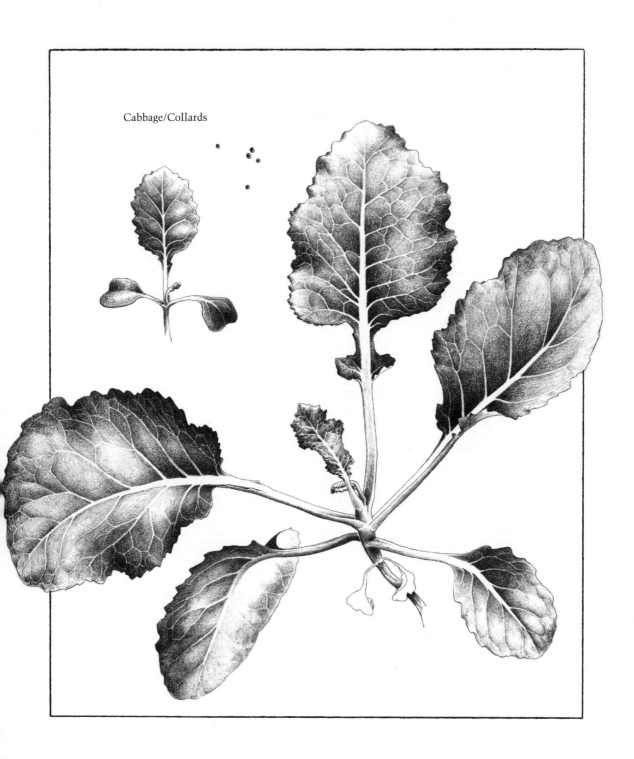

Cabbage/Collards

Broccoli

Cruciferae
  Brassica oleracea var. *italica*

Broccoli takes two distinct forms. The type most of us think of when broccoli is mentioned is the "green sprouting" or "Italian" broccoli, which is illustrated opposite. More popular in Europe is the type known as "winter cauliflower," "heading broccoli," or "cauliflower broccoli." Most confusing. The European form at harvest looks like the white heads of our familiar cauliflower; it is a hardier but slower growing plant than our common cauliflower. Broccoli forms small side shoots after the main head is cut; true cauliflower plants do not. Some sources list cauliflower and broccoli under the same classification as essentially the same plant, *B. oleracea* var. *botrytis.*

Green heading broccoli is a more recent arrival to our gardens than most of our other vegetables and herbs. It was developed from the cabbage-type plants as a distinct variety only about two thousand years ago. It was quite unknown in this country until the early 1900s when it was popularized by Italian immigrants.

Fortunately, our common forms of broccoli and cauliflower are not too difficult to tell apart once they have a leaf or two about them. They begin with the same two cotyledons and smooth blue-green leaves as the rest of those sharing this genus and species name. Broccoli quickly develops prominent "finger" shapes at the base of the leaves; our common white-heading cauliflower has leaves that are a longer, flatter spear shape by comparison, with a less complicated leaf outline.

Broccoli leaves in shape look most like those of the turnip, rutabaga, and radish. The texture of the leaves of these three species, however, is quite different. Broccoli is smooth and hairless, with the central stem elongated, whereas turnip, rutabaga, and radish leaves are greener, rough and bristly, and grow upward from the base rather than from a central stem.

The less common plant known as broccoli raab, *Brassica ruvo,* is more closely akin to the turnip. The leaves of this vegetable are smoother than the turnip's but have the turnip leaf's shape. Unlike the turnip, the broccoli raab grows upright from an elongated stem. It is the tender leaf buds and shoots that are eaten.

Broccoli, a cool-weather plant, should appear about ten days after sowing.

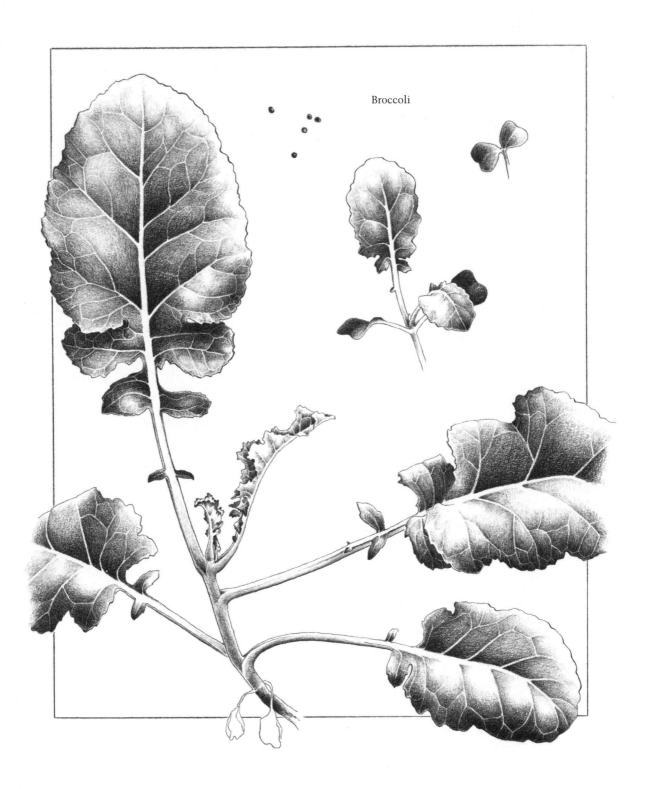

Broccoli

Cauliflower

Cruciferae
  Brassica oleracea var. *botrytis*

White heading cauliflower has much the same history as green sprouting broccoli and the other vegetables derived from field cabbages first cultivated in Asia Minor and the Mediterranean region. The oldest written records of cauliflower as a distinct vegetable crop date back to the sixth century B.C. when primitive forms were grown in the Near East. Varieties adapted to the cooler climates of northern Europe were developed during the Middle Ages, and it is from these cool-weather plants that our modern varieties were selected and further refined.

Cauliflower has the same two heart-shaped cotyledons possessed by the rest of the family *Cruciferae*. The seedlings bear a strong resemblance to their close cabbage cousins—both have similar smooth, bluish leaves. Most early seedling confusion occurs between the cauliflower, broccoli, and green kohlrabi. These all have fairly flat, spear-shaped leaves with fine toothing on the edges. Broccoli develops more complicated and irregular "finger" shapes at the base of its leaves compared to the cauliflower, and the broccoli's leaves are slightly buckled. These differences become more pronounced by the fourth and fifth leaves. Kohlrabi leaves are distinctly more pointed, even at the earliest stages.

Cauliflower enjoys cool temperatures and germination proceeds as rapidly as with other close relatives of the *Brassica* genus, with seedlings appearing in seven to ten days.

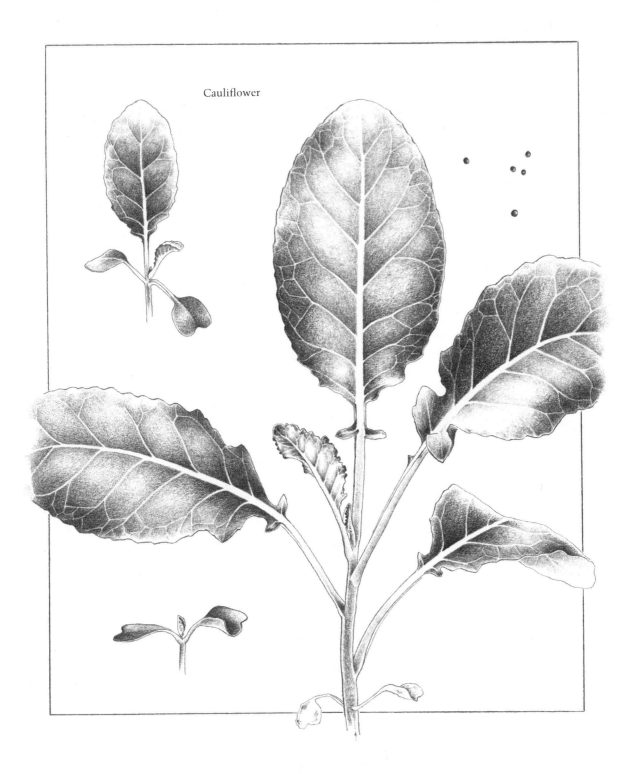

Cauliflower

Kohlrabi

Cruciferae
    Brassica oleracea var. *caulorapa (var. gongylodes)*

Kohlrabi is a German word meaning "cabbage turnip." This is a descriptive term rather than a botanical explanation; it refers to its development from the Mediterranean field cabbage and the resemblance of its edible enlarged stem to the turnip root. Although it is still a matter of some debate, kohlrabi is probably a comparatively recent arrival to cultivation. It was apparently virtually unknown anywhere until about four hundred years ago when it was first described in northern Europe, although it is possible that similar forms were cultivated to some extent by the Romans at a much earlier date. By the end of the sixteenth century it was well-known throughout Europe, Spain, Italy, and the eastern Mediterranean countries. It was introduced to North America in the early 1800s.

There are a half dozen or so cultivars of kohlrabi offered now in this country. Their color is commonly purple, whitish, or yellow-green. The purple varieties are quite unmistakable for the color which affects all parts of the plant, stem, veins, and leaf. Initial confusion may occur with varieties of purple cabbage but, on closer examination, cabbage leaves are always much rounder.

White or green cultivars of kohlrabi look distressingly like several of their close cousins as seedlings. The resemblance begins with the appearance of the usual pair of heart-shaped cotyledons followed by fairly thick, smooth, bluish-green leaves. Even those first kohlrabi leaves on close examination are seen to be much more jagged in outline, more sharply pointed, and longer than those of the cabbages, collards, Brussels sprouts, cauliflower, or broccoli. They perhaps most resemble the leaves of some of the kale varieties. The most common forms of kale have very frilly leaves that are not flat like the kohlrabi's but curl at the edge. The kohlrabi stem is short, and the thickening stem is not apparent for at least a month after the seedling's initial appearance.

Kohlrabi seedlings should appear in about twelve days.

Kohlrabi

Brussels Sprout

Cruciferae
  Brassica oleracea var. *gemmifera*

Brussels sprouts are relatively recent arrivals to our gardens and tables. Along with cabbages, collards, kale, broccoli, kohlrabi, and cauliflower, Brussels sprouts count the wild field cabbages of the Mediterranean as their ancient predecessors. It is likely that this plant did indeed evolve in or near Brussels, Belgium. It has been cultivated as a distinct botanical variety for about four hundred years; the first written description of this vegetable is dated 1587. Brussels sprouts have been grown in the United States only since the early 1800s.

The Brussels sprout is an odd-looking plant at harvest time but its development from the cabbage is easily demonstrated. If the head is cut from a common cabbage, often numerous smaller heads will sprout from the remaining stem. The Brussels sprout is really a more refined, long-stemmed form of cabbage.

Its heart-shaped cotyledons easily mark the Brussels sprout as a member of the *Cruciferae* family: their bluish, greenish tinge indicates their close relationship with the cabbages. Brussels sprout seedlings bear closest resemblance to the green heading cabbages, collards, and cauliflower, and to a lesser extent broccoli and Chinese mustard. All of these have a similar round, slightly scalloped leaf shape.

One distinguishing feature of the Brussels sprout is its habit of cupping its roundish leaves backward. Collard, cabbage, and cauliflower leaves are flat by comparison. The cabbage and collards have heavily and irregularly scalloped leaves; cauliflower leaves are longer and spear-shaped.

The broccoli leaf is also slightly buckled but more complex than the silhouette of the Brussels sprout leaf.

Chinese mustard plants have spoon-shaped leaves but they differ from the Brussels sprout seedlings in that their leaves are unmistakably green, not bluish, and they have very white thick stems.

Brussels sprout seedlings should appear in about ten days. Having been developed in Northern Europe, this plant is very hardy and may over-winter in the garden, flowering the following spring.

Brussels Sprout

Turnip

Cruciferae
  Brassica campestris var. *rapa*

It is not definitely known where turnips originated. Russia, Siberia, and the Scandinavian peninsula have been suggested, as primitive wild forms can be found there. Wherever it came from and was first developed, the two basic types of turnip known today were widely cultivated in ancient times. One type is savored for its leaves, and one is grown for its enlarged root.

The turnip was first brought to North America by the explorer Jacques Cartier, who planted it in Canada in 1541. Seeds were brought to Virginia and Massachusetts by the first English colonists.

The turnip's membership in the family *Cruciferae* is obvious, given its initial pair of heart-shaped cotyledons. The first true leaf when fully unfolded bears some likeness in outline to cabbage, collard, cauliflower, and Brussels sprout seedlings. Although all of the above-named vegetables have round, slightly scallop-edged first leaves similar to the turnip's, their leaves are very smooth and blue-green. The turnip leaf is green, somewhat hairy, and rough in texture.

Radish and rutabaga seedlings are more apt to be confused with turnip because their leaf surface is similar in texture and color. The radish seedling will soon distinguish itself by its narrower and rapidly more complex leaf outline; many more finger-shaped ins and outs appear at the base of the leaf. An additional difference is that most of our common radishes have distinctly red stems, in contrast to the turnip's pinkish one.

The rutabaga was probably derived from an unusual cross-pollination of two quite different plants—a cabbage and a turnip. Like the turnip, the rutabaga usually has a pinkish stem. Rutabaga leaves are slightly less hairy, rounder, and with a slightly blue tinge to the green. The seedlings are difficult to tell apart.

Both types of turnip, those grown for greens and those grown for roots, will look identical as seedlings. They should appear about seven days from planting.

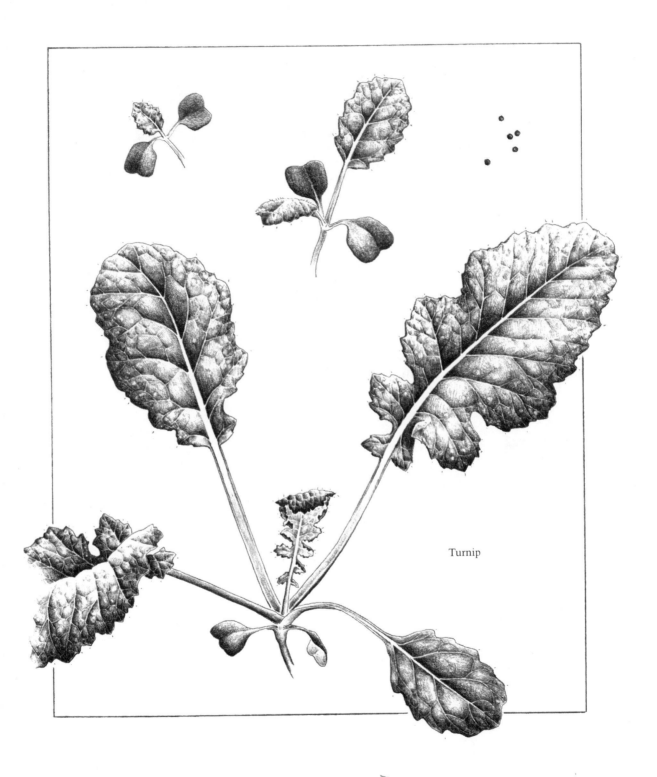

Turnip

Rutabaga

Cruciferae
  Brassica napus var. *napobrassica (Brassica campestris* var. *napobrassica)*

Rutabagas are also commonly known as Swede or Swedish turnip. Botanically, they have been known by a confusing array of names. All of the names given them indicate their strong ties to the turnip; their differences lie mainly in their genetic makeup. It is generally believed that the rutabaga is the result of an unusual hybridizing of the 18-chromosome cabbage with the 20-chromosome turnip—giving us the 38-chromosome rutabaga. This new species was first mentioned in the late Middle Ages in northern Europe, and it may have Scandinavian origins.

As further evidence for the cabbage/turnip combination theory, the rutabaga seedling resembles both the cabbage-related seedlings and the young turnip plant. The very first rutabaga leaf to unfurl looks quite like the cabbage or collard leaf. The following leaves look more like those of the broccoli in outline, yet they are just a little more tender, a little rougher, ever so slightly hairy. Then too, all the leaves emanate from the base rather than from an elongated stem like the broccoli. Compared to the turnip, the rutabaga is blue-green in color and a bit rounder in leaf shape. This is more obvious when the seedlings have grown to four or five leaves.

The rutabaga requires a longer growing season than the turnip and also germinates a little more slowly. It usually takes nine days for the seedlings to appear.

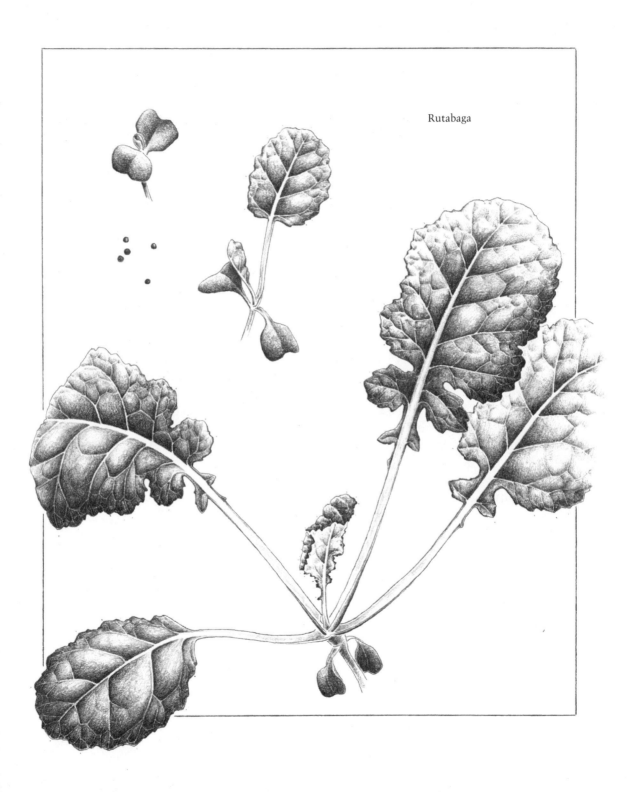

Rutabaga

Mustard

Cruciferae
  Brassica juncea

Indian mustard, or leaf mustard, is usually called simply "mustard" in this country. It is of the same family and genus as Chinese mustard, but of a different species. The mustards that are grown for their seeds are also different species, *Brassica alba* and *B. nigra*. These are not normally grown in home gardens.

Indian mustard evolved in several stages at least two thousand years ago in central Asia. It seems to have spread more rapidly than its cousins the oriental cabbages and mustards, perhaps because it was more easily adapted to other climates.

Mustard has cotyledons identical to those of its fellow members in the family *Cruciferae*—a pair of dark, vaguely heart-shaped leaves. One must wait until the next leaves completely unfold for further identification. The most common varieties have deeply scalloped leaf edges and very recessed veins which make them curl and dimple; this characteristic becomes more pronounced with each leaf to follow. The leaves are covered with fine hairs. The seedling illustrated is the popular cultivar, "Green Wave." Some confusion might be caused by the turnips and rutabagas which also have green, fairly rough-looking leaves.

Different varieties of mustard have differently shaped leaves. The less common cultivars, variously called "mustard cabbage," "mustard spinach," or "Tendergreen," have broader, smoother leaves than those depicted here. The leaves of all varieties are a deep green color that easily separates them from the large group of cabbage-related plants with smooth bluish leaves, which they otherwise initially resemble.

Mustard seedlings should appear about nine days after sowing.

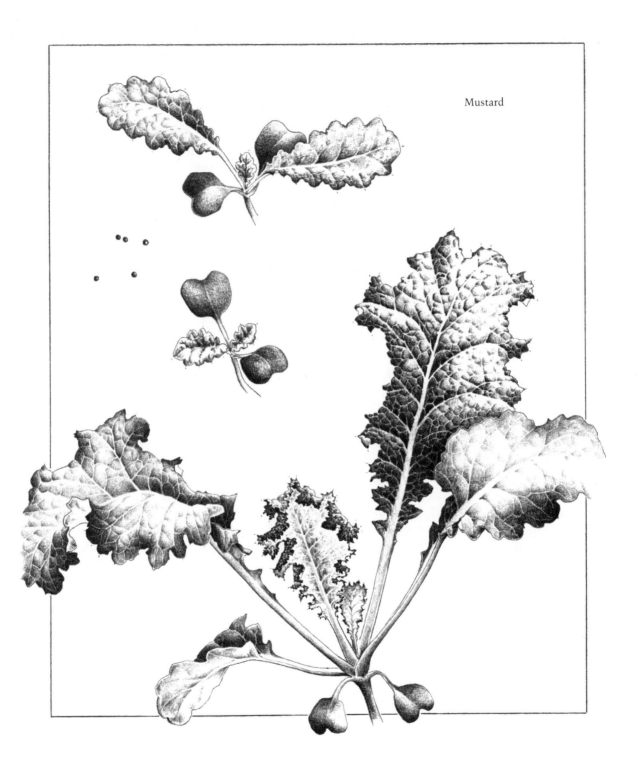

Mustard

Chinese Cabbage

Chinese Mustard

Cruciferae
  Brassica pekinensis
  (B campestris var. *pekinensis*)

Cruciferae
  Brassica chinensis
  (B campestris var. *chinensis*)

Both of these oriental vegetables have been an important part of the diet of the people of eastern Asia, China, and Japan since time immemorial. These civilizations have selected and developed such an amazing diversity of varieties that making distinctions between them, and thus naming these plants accurately, is sometimes quite difficult. Chinese cabbage and Chinese mustard have been introduced only relatively recently to foreign climates. Their first recorded appearance in Europe was in 1751, but it was not until the late 1800s that they were viewed as anything more than curiosities for the botanists' gardens.

Chinese cabbage usually has leafy, compact heads with a thick, white, central vein. This is sometimes also called celery cabbage or pe-tsai. The illustration shows a popular cultivar, "Michihli." Chinese mustard, also frequently called pak-choi, is gaining acceptance in North America as well. This plant has smoother, spoon-shaped leaves with thick white leaf stems.

Both varieties, Chinese cabbage and Chinese mustard, have identical pairs of cotyledons—the same heart-shaped first leaves that belong to the entire *Cruciferae* family. Once the first true leaf has opened out completely, identification becomes possible: the leaves of both of these oriental favorites are very green, which is an immediate hint that they are not of the bluish-greenish *Brassica oleracea* cabbage clan.

The most common Chinese cabbages have spear-shaped leaves. These are similar in texture to the rough, bristly leaves of the turnip, rutabaga, or radish. Chinese cabbage leaves gradually widen out from shortened leaf stalks; the leaf outline is simple, the edges heavily toothed.

Chinese mustard leaves are quite different from those of the Chinese cabbage. These smooth, green leaves are round and buckled with long, round, white leaf stalks. The mature Chinese mustard plant looks somewhat like our common Swiss chard.

Both of these Oriental vegetables should appear about nine days after sowing.

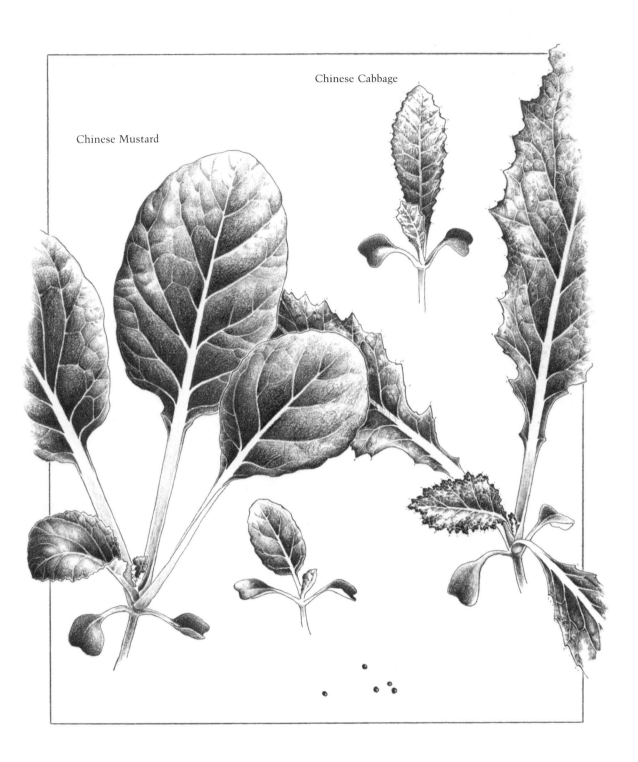

Chinese Cabbage

Chinese Mustard

Radish

Cruciferae
  Raphanus sativus

Radishes have had a long history of cultivation in China and in Central Asia, their probable homeland. Egyptian records show that radishes were a common food there even before the pyramids were built. Greeks and Romans of the third century B.C. also favored this vegetable.

There is an enormous variety of shapes and sizes and flavors of radishes not yet explored by North American gardeners. Some oriental varieties are grown for their edible seed pods, the oils derived from the seeds, or for their leaves, which are cooked and eaten. The small red and red-and-white garden varieties that are popular here now were not documented until the mid 1500s. Radishes can be black-skinned, white, or rose-colored. They can range from thimble to basketball size, and some cultivars grow to two feet long. Varieties have been developed that can over-winter in the soil to be harvested in spring.

The genus name *Raphanus* is a latinized form of an old Greek expression meaning "easily reared." Radishes grow so fast that there is little time to wonder about their identity before it is time to harvest the bulb. The two cotyledons of the garden radish are similar in shape to, but usually much larger than, those of the other vegetable plants in this family. Their leaves are a very green color, easily distinguishing them from the blue-green leaves of the cabbage-related genus *Brassica*.

In shape, radish leaves most closely resemble those of the turnip, rutabaga, and broccoli. The broccoli seedling, however, has the aforementioned coloring of the cabbage clan, its leaves smooth compared to the radish seedling's. Turnip and rutabaga seedlings are more easily confused with the radish as their leaves share the same slightly bumpy and bristled texture. Another characteristic these three share is that their leaves all grow out from the base rather than from an elongated stem.

Notice that the radish leaves are more slender and rapidly become much more complicated than either turnip or rutabaga leaves. Our most common varieties of radish have stems that are bright red from their first appearance. Turnip and rutabaga stems are pale pink at best.

Radish seedlings appear quickly, usually within six days.

Radish

Common Bean

Leguminosae
  Phaseolus vulgaris

The scientific name *Phaseolus vulgaris* encompasses a wide variety of beans with kidney-shaped seeds, including not only the edible-pod beans (green string or snap beans and wax beans), but also the dry shelling types such as navy or pea beans, red kidney, marrow, and pinto beans. Lima beans, *Phaseolus lunatus*, were so-called because Lima, Peru is one place where early explorers first tasted them. Less familiar to North American gardeners are the scarlet and white runner beans, *Phaseolus coccineus* (or *P. multiflorus*), highly ornamental pole beans. All of these species originated in Central and South America and Mexico and are vegetables of great antiquity. Their use had spread extensively throughout the Americas by the time of the first European explorations, and they were thus brought back to the Old World countries in the 1500s.

The illustration opposite is of the most popular form of garden bean—the green bush bean. All of the many types of shell beans and wax beans have seedlings similar to these. They are unselfconscious seedlings, large and unlike any weed. Their cotyledons are thick and do not function as leaves but contain stored food for the new seedling; they soon shrivel and fall. A pair of simple, pointed leaves soon appears, followed by triangular leaflets in groups of three.

Lima beans, belonging to the same genus but a different species, will also look quite like green bean seedlings. On closer observation, however, their cotyledons are broader, and the coloration of the first pair of leaves of most lima varieties will be a darker green, with a lighter green/gray mottling along the major veins. Lima bean leaves are perhaps a bit longer and more sharply pointed. These, unfortunately, are not highly reliable characteristics; if some confusion occurs, resign yourself to waiting until the pods form for precise identification.

Mature runner beans are similar to other pole-type beans but their unusual behavior in their early stages sets them apart. Runner beans leave those thick cotyledons behind under the soil when the first shoots uncoil. This generally gives the first folded leaves a dirtier arrival into the world. Once unfolded, these leaves resemble those of the other beans.

All of these bean seeds take offense at being planted in cold, wet weather, and may rot if so treated. Under reasonable conditions they should appear speedily in about six days.

Common Bean

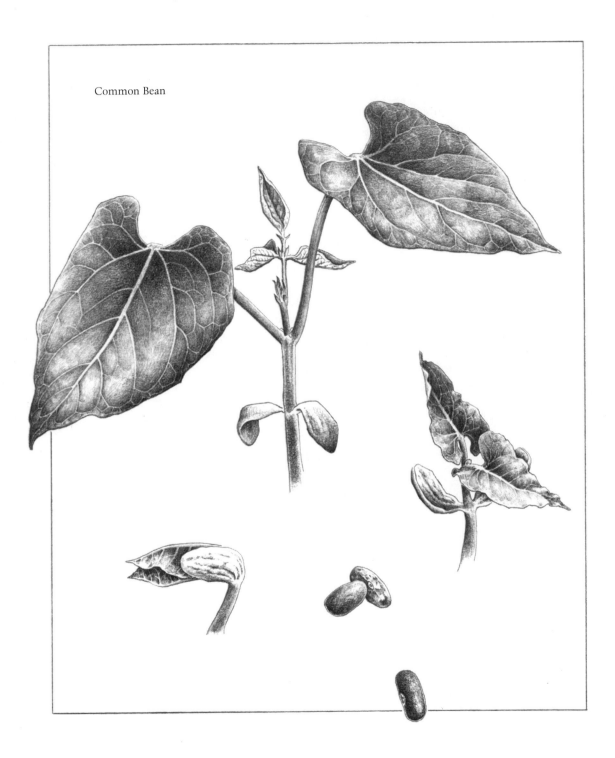

Garden Pea

Leguminosae
  Pisum sativum

Peas have been known as an important source of protein since ancient times. The main locus of origin and domestication was probably Central Asia, with secondary developments in the Near East and North Africa. Our popular green shell peas most likely were derived from the gray-seeded field peas that were used in dried form. It was not until the twelfth century at the earliest that people began to shell peas at an immature stage and eat them green. They became quite the rage for the privileged class in France by the 1500s. Edible-podded peas, *P. sativum* var. *macrocarpon*, were also known at that time. This variety is sometimes called "snow peas" or "Chinese peas," although they were not developed in the Orient. Green shelling peas did not become truly common until the eighteenth century.

Peas have either wrinkled or smooth seeds, depending on the cultivar. All peas leave their two cotyledons under the soil and emerge with a cluster of folded leaves. The leaves of all of our common peas are smooth, hairless, round, and arranged in pairs. Tendrils soon appear. It is impossible to tell at the seedling stage whether an adolescent pea will become a "snow pea," bush pea, or one of the newest cultivars, the "sugar snap pea."

All pea varieties are quite tolerant of cool weather, and the seeds can even be planted before the last frost. They do germinate more quickly in warmer soil, though, and can appear in as few as seven days.

Garden Pea

Cowpea

Leguminosae
   Vigna sinensis var. *unguiculata*

There are many descriptive names for the cowpea. Some cultivars are known as "black-eyed peas" for the distinctive coloration of the seeds. Others are called "crowders" because the seeds are quite crowded in the long thin pods. "Southern pea" is yet another popular name attributable to the cowpea's popularity in the warmer states.

Most geobotanists consider India to be the land of origin of the cowpea, but there is some uncertainty due to the plant's great antiquity and unrecorded early wanderings. In India, the cowpea has many names in ancient languages, including Sanskrit, indicating that it was in cultivation there in prehistoric times. One theory has it that this legume was carried from India to Arabia and Asia Minor; from there its popularity spread to Central Africa, where it can be found growing wild.

The cowpea reached the Americas from Africa via the slave traders, who discovered that it was a cheap source of food for their cargo. It was first introduced to Jamaica and adjacent islands and from there came to Florida around 1700.

In form, the cowpea seems to have more in common with our garden bean than it does with our usual pea plant; in fact, the names "bean" and "pea" are conferred upon plants largely by custom rather than for technical reasons. All beans and peas are included in the same family of plants.

Like the common bean's, the cowpea's first appearance is rather obvious. It is a large seedling whose thick elbow of a stem shows first. This is followed by a pair of large, fat cotyledons. The cowpea continues to copy the common bean's pattern by then unfolding a pair of simple, dully pointed leaves, and next come the usual triangular bean leaves in sets of three. The identifying feature of the cowpea is that its stem is squarish rather than round and its leaves are darker green, smooth, and shiny as compared to the bean's faintly fuzzy leaves.

Cowpeas germinate a bit slower than green beans. They will appear in about ten days if planted in warm soil.

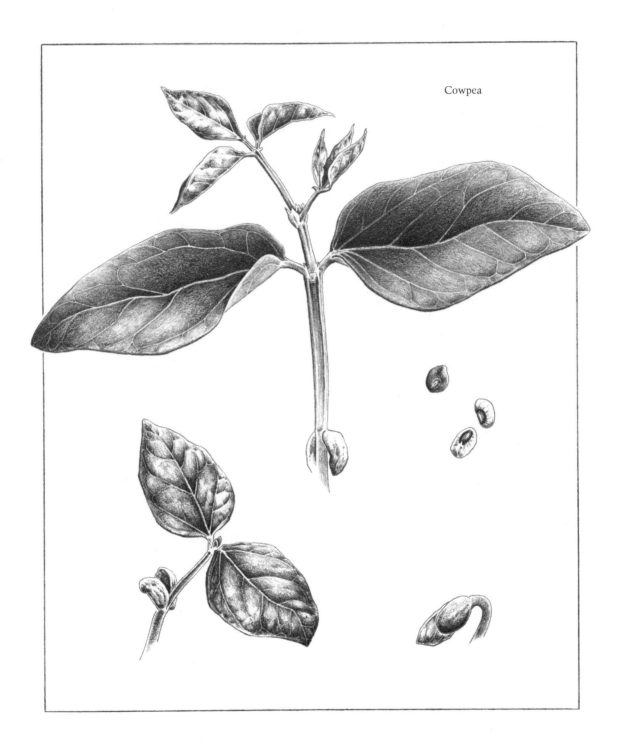

Cowpea

Fava Bean

Leguminosae
  Vicia faba

The fava or broad bean is little known in North America, but it was one of the first plants to be actively cultivated by mankind. Different sources list its beginnings in Northern Africa and the Near East, or Central Asia and the Mediterranean. Wherever it was first deliberately singled out and planted in an organized fashion, the fava was the only important bean eaten in the Old World until the discovery of the now more familiar kidney-type beans of Central America and Mexico.

The fava bean is a large seedling; the seed itself is enormous compared to other seeds of this family. Its first folded leaf appearance is like that of the common pea: a sudden eruption of a cluster of leaves, rather than a slow unfolding from a pair of visible cotyledons, as with our common beans. The fava bean supports its pairs of thick, round leaves with an upright, square-sided stem. Its tendrils are small, almost non-existent.

This unfamiliar bean prefers a long, cool growing season and can even withstand a light frost. The seedlings should appear in about seven days.

Fava Bean

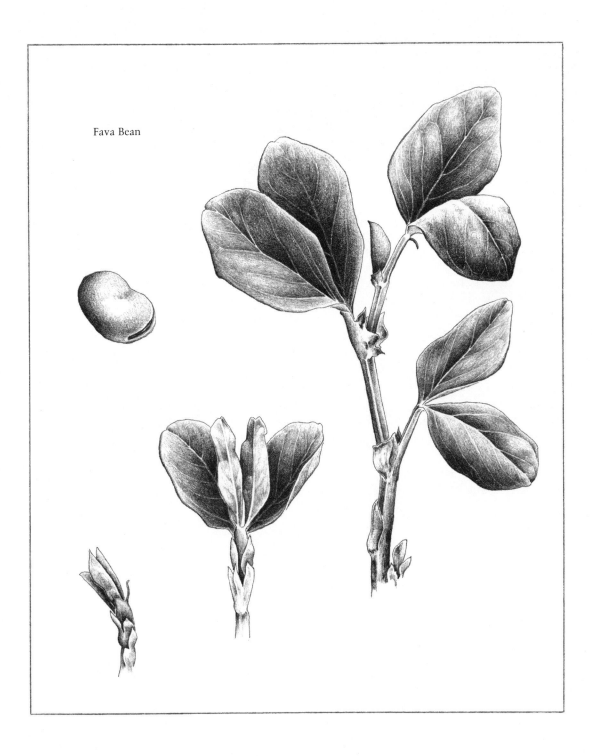

Lentil   Chickpea

Leguminosae   *Leguminosae*
  Lens esculenta   *Cicer arietinum*

Both of these plants are believed to be indigenous to western Asia and have been used since ancient times. Lentils are certainly one of the oldest of the legumes to be actively cultivated, having been introduced into Egypt, Greece, and as far east as China well before Biblical times. The chickpea or garbanzo is also a plant of ancient use. The wild forms from which it was derived have either disappeared or this vegetable has changed radically under cultivation, making it difficult to decipher just how it was developed.

Neither of these legumes is commonly grown in backyard gardens. Like most dry bean/peas they seem to require more time and space than most people are willing to allow them. Those who do experiment with these protein-rich vegetables may be hard pressed to recognize them from the weed plants.

Whether they are more properly called beans or peas is quite irrelevant, but their initial appearance is similar to the emergence of the common green pea in that both leave their cotyledons in the soil. The first small, oval leaflets of the lentil and chickpea are tightly folded.

Chickpeas on close examination have fine silver hairs on leaf and stem. The leaves will always appear in odd numbers on each leaf stem. It is the larger of the two plants, but not by much.

Lentils have thin, often reddish, stems and smooth, creased leaves. These leaflets will always appear in even numbers.

Both of these seedlings of unimpressive size resemble some of their wild cousins, particularly the vetches, the partridge pea, and some of the trailing clovers.

The seedlings of both plants appear in five to eight days in warm weather.

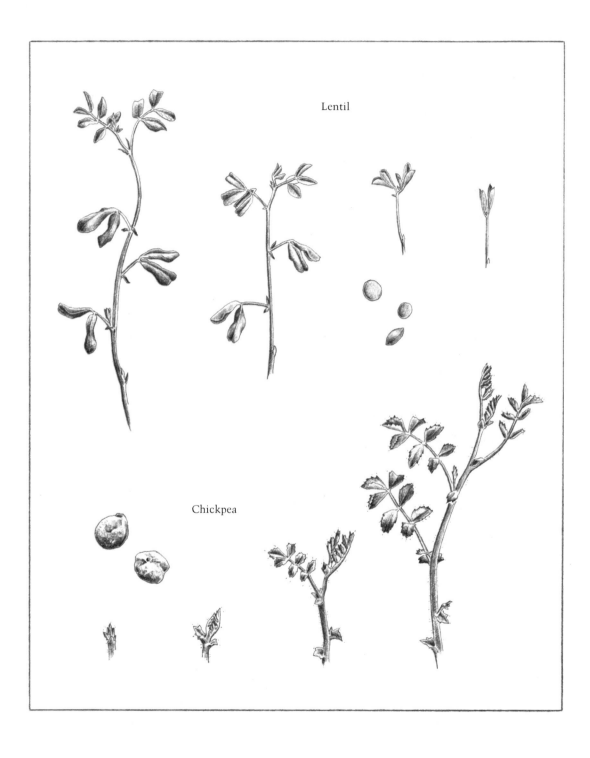

Lentil

Chickpea

Soybean

Leguminosae
  Glycine max (Glycine soja)

The soybean is a relative newcomer to the Western world but one of the most ancient and important foods in the Orient. Its origins are in China, where it has been grown for more than five thousand years. The first kinds to be cultivated were dry field beans used for flour, oil, and fermenting; the green "vegetable" varieties that are shelled and eaten green were a later development. Field soybeans used for stock feed and soil building were introduced to the United States somewhat earlier, but the vegetable soybeans that are of primary interest to home gardeners have only been offered for sale here for about thirty years.

The soybeans' first appearance is after the fashion of common garden beans. The thick, round cotyledons are dark and shiny when compared to those of the green bean. The first pair of true leaves to appear are round, and the next leaflets continue in the same pattern that garden beans follow, in groups of three. The leaves, stems, and even the seed pods of most varieties of vegetable soybeans are covered with a fine, downy hair. The whole plant is vaguely reminiscent of a gigantic clover plant because of its soft, round leaves. Clover is in fact a fellow member in the family *Leguminosae*.

Soybean seeds do not usually remain viable for longer than two years. Seedlings should appear in about twelve days.

Soybean

Peanut

Leguminosae
  Arachis hypogaea

The peanut began its involvement with agriculture in South America. It has been found in early Peruvian tombs, indicating that it was under cultivation before the first century A.D. It has changed so much over time that the wild forms from which it evolved are not known. Portuguese explorers of what is now Brazil apparently brought the peanut to East Africa. A century later it came back to the American continent with African slaves. Peanuts are also known as groundnuts or goobers, an African word for groundnuts.

Peanuts have rather odd flowering habits. The first flowers to appear are bright yellow and quite showy; they are, however, sterile and so do not produce seeds. Secondary, less-conspicuous flowers appear in the lower leaf axils. These flowers are fertilized and form peduncles, also called "pegs" or runners, which bend over and bury themselves in the soil. The peanut seed forms underground at the end of these runners.

Yet another curious characteristic of the peanut plant is that its leaves fold tightly together at night and open again during the day.

The peanut, as its name suggests, is related to the pea/bean family of legumes. Its first appearance is like that of the pea—an abrupt cluster of folded leaves. These leaves as they unfold are smooth and round and array themselves in two pairs on each leaf stem. Quite often the two cotyledons are visible just at the surface of the soil. Clover, a common weed plant which is also of this family, is similar in form to the peanut but is much smaller.

Peanut seeds can be planted with or without their shells intact, but removing the shell will hasten germination. If the soil is warm, seedlings should emerge in about seven days. If the weather is cloudy or cold, it may take twice as long. The seed does not remain viable for long and should be purchased fresh each year.

Peanut

Okra

Malvaceae
  Hibiscus esculentus

Okra is also known in this country as "gumbo." Both words are corruptions of African words for the plant. Okra seems to have originated in and around what is now Ethiopia, and was sown and harvested several thousands of years ago. French colonists introduced okra to the Mississippi delta region of the United States in the early 1700s. This very ornamental vegetable with its attractive leaves and showy flowers can be grown over most of North America, but it is not as yet widely popular.

The two green cotyledons of the okra are a bit crumpled when they first appear. They look like a smaller version of the wide, round cotyledons of the melons and squashes, although they are not related. If you are accustomed to seeing only mature okra plants, the seedlings might surprise you. The older plants have a complicated, radiating five-fingered outline; the young seedlings have broad, simple leaves and look rather like incipient maple trees. The okra is a sturdy, upright seedling. The base of the leaves and leaf stems of most varieties of okra are a striking bright red color.

Okra prefers a warm climate, especially to get started—its seeds will rot in a cold, damp soil. These seedlings should appear in about ten days if they are sown after the soil has warmed. Presoaking the seeds overnight may help them appear even more quickly.

Okra

Carrot

Umbelliferae
   Daucus carota (Daucus carota var. *sativa)*

The carrot is believed to have originated in Asia Minor or the Near East. It has been known since ancient times but its first uses were medicinal rather than culinary. The popularity of the carrot spread quickly and it was certainly cultivated in the Mediterranean region long before the beginnings of Christianity. By the thirteenth century carrots were being grown in southern Europe and China; they were introduced to the Americas in the late 1500s and early 1600s by European explorers. It has only been within the last hundred years that the flavor of carrots has improved to the extent that they are now popularly eaten raw.

The first cultivated carrots no doubt had the much smaller roots that were an advantage to the wild carrot. The civilized carrot with its large single root is much more fragile—easily uprooted and subject to drought—and with its higher water content, freezing has disastrous effects. It is interesting to note that the garden carrot and the Queen Anne's Lace, a common introduced weed, share both genus and species name. This delicate weed was probably one of the carrot's ancestors and they will readily cross-pollinate.

The important cultivars of garden carrots in this country are orange and of various cylindrical shapes. There is a remarkable diversity in shapes and colors throughout the world, though, ranging from purple to yellow and white, and in Japan there grows a variety which commonly reaches three feet in length.

The diminutive carrot seedling is always at risk of being pulled up with the weeds. It has a pair of rather nondescript cotyledons which resemble a variety of weeds and, of course, all of its relations in the family *Umbelliferae.*

As the first leaves unfold, the carrot's closest look-alike among the vegetables and herbs is the caraway. The caraway seedling is a smaller, slower-growing plant whose leaves are a little broader than the finely-cut carrot leaves. Dill and fennel are also similar but much more delicate; their leaves are threadlike compared to the carrot and caraway.

Carrot seedlings should appear about eight days from sowing. Carrot and dill are said to have a pronounced dislike for each other and should not be planted together.

Carrot

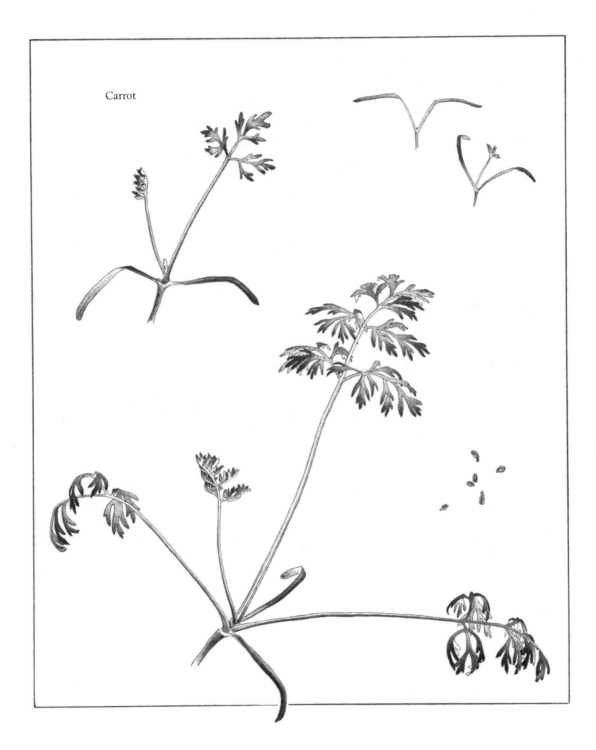

Celery

Umbelliferae
   Apium graveolens (var. dulce)

Primitive forms of celery have been used medicinally and as a flavoring for a long time, but celery as we know it today was not popularly eaten until the late eighteenth century. Wild celery grows in swampy areas in Europe, Asia Minor, the Caucasus, southeastward to the Himalayas, and in the Mediterranean lands where it may have begun. The wild plant is much like the stockier plant known as smallage. Improvements came gradually, as celery first began to be cultivated as an herb in the gardens of Italy and northern Europe in the 1500s. By the beginning of the 1600s celery began to be eaten as a vegetable, but it was not until the late 1800s that celery had improved to the point that blanching and over-wintering were not considered essential to improve its flavor.

Celery is a painfully small seedling. When it finally appears, just as you've given up hope, it looks like a dozen other tiny two-leafed weed plants. This first pair of leaves is light green and slightly pointed; the first real leaf comes up ever so slowly. This leaf, and the ones that follow, look like smaller versions of parsnip, to which celery is related. To a lesser extent it also resembles parsley, caraway, and coriander seedlings.

   Celeriac, *A. graveolens* var. *rapaceum,* is less common in North American gardens. It was developed from the same wild marsh plants as the celery, at about the same time. In this case, it is the thickened round root that is harvested. As a seedling it looks and behaves much like its celery cousin.

Celery is a most difficult seedling. To begin with, the seeds prefer a cooler temperature than most other vegetable seeds—60–70 degrees F.; they will refuse to germinate at a temperature of 85 degrees or more. Fluctuating day/night temperatures will aid in their germination. Even if you have done all you can to please them the seedlings may still take three weeks or longer to appear.

Celery

Parsnip

Umbelliferae
  Pastinaca sativa

Parsnips are believed to be native to the eastern Mediterranean area and north-eastward through Europe. They were known to the early Greeks and Romans, who used to harvest them from the wild. It is unknown where or when or who first began to study the wild parsnip and to improve the size and flavor of the root. By the mid-sixteenth century parsnips were a common staple in Europe and elsewhere, and were introduced to North America with the first colonists.

This vegetable is slow to germinate and slow-growing as well. It begins with an unimaginative pair of cotyledons which at first resembles many of our common weeds, as well as the cotyledons of some of its close vegetable and herb relatives—celery, parsley, and coriander.

When the first leaf has formed, its strongest resemblance is to celery and parsley. The parsnip does indeed look like an enormous celery seedling. Its stem is whitish, as would be expected; its leaves, which all grow upward from the crown, are basically round with scalloped edges. As the plant matures, the leaves become much more complex—each leaf divided into many leaflets, the leaflets themselves divided and scalloped. If left in the ground over winter parsnips will send up a tall flower stalk in spring.

Parsnips are quite hardy and may persist as weeds if allowed to go to seed. Studies show that parsnip seed does not remain viable for more than a year or two. Seedlings should appear in about eighteen days.

Parsnip

Florence Fennel

Sweet Fennel

Umbelliferae
  Foeniculum vulgare var. *dulce*

Umbelliferae
  Foeniculum vulgare

Various forms of fennel have been used from time to time in the past as food, flavoring, medicine, ritual charms, beverage, and insect repellant. Fennels are probably native to the Mediterranean lands and part of Asia. Nowadays, three major varieties of fennel are cultivated. Sweet fennel is found in the herb garden. Its leaves and seeds are used mainly for flavoring and for tea. Florence fennel or finocchio is grown for its thickened basal leaf stalks. A third variety less common here is carosella, or Sicilian fennel; it is harvested for its young stalks, which are eaten like celery.

The illustration is of Florence fennel. The other varieties are identical to this as seedlings; only after several months will it begin to be possible to distinguish between them. The Florence fennel will have obviously larger stems which overlap at the base, carosella will have smaller stems. The herb fennel has a longer main stem, very slender stalks, and taller growth.

As young seedlings, fennels and dill, members of the same family, are all but impossible to tell apart. Both have long, narrow cotyledons followed by delicate feathery leaves. The fennel will be almost imperceptibly larger with more thread-like foliage.

Fennel seedlings also look to some extent like young carrot or caraway plants. By comparison these last two plants have flatter and coarser-cut leaves; fennel and dill have a stronger scent.

Fennels reportedly do not get along well with many other vegetables and herbs. Most often mentioned is an antipathy toward bush beans, caraway, tomatoes, and kohlrabi. Fennel seedlings should emerge in twelve days.

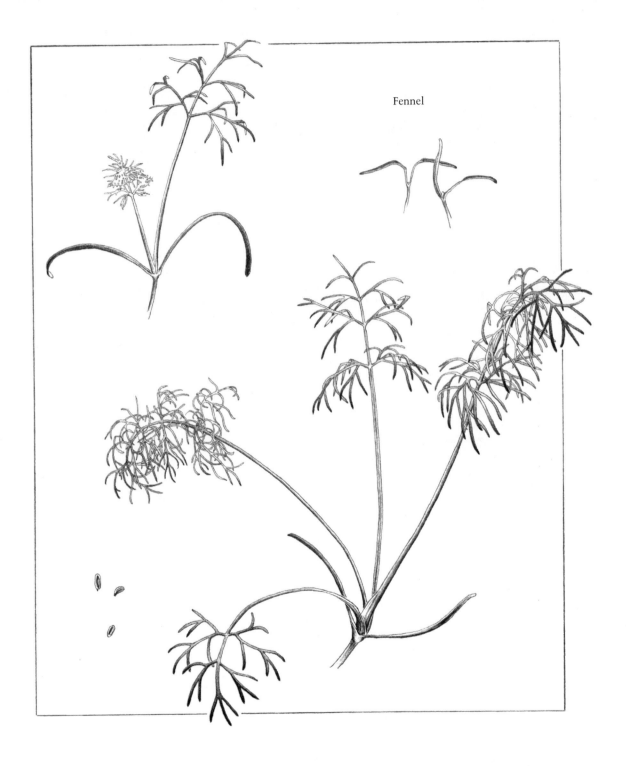

Fennel

Dill

Umbelliferae
  Anethum graveolens

Dill is a hardy annual herb grown for its seeds and leaves. It is believed to be native to the lands of the Mediterranean and the Black Sea. It is found today growing wild in many parts of Europe and Asia. Although it was grown for medicinal and flavoring purposes by many of the oldest civilizations, it did not reach the British Isles until the sixteenth century and was introduced to North America by way of England with the arrival of the colonists.

Most seedling confusion occurs between dill and fennel. The two are virtually indistinguishable at these early stages; both have long narrow cotyledons similar to the carrot and caraway. The next leaves are lace-like fronds on smooth, hollow stems. Fennel seedlings are generally ever so slightly larger and more threadlike than the dill. Most forms of fennel will later show thicker stalks rising from an overlapping base. The dill will tend to have one main stalk that rises and eventually branches. Carrot and caraway leaves are flatter and not as finely-cut.

Dill has an adverse effect on carrots—do not plant them together. Seedlings should appear two weeks after sowing. Dill seed is winter-hardy and this plant will readily resow itself.

Dill

Caraway

Umbelliferae
  Carum carvi

Caraway is most likely indigenous to Europe although its use in Asia, the Mediterranean, and the Orient also predates recorded history, making an accurate guess as to its origins difficult. When we think of caraway now it is the seeds that come to mind, but the leaves and yellowish, carrot-like roots can also be eaten. There are different varieties of caraway—some are annuals and some are biennials. For either type, germination and early growth is quite slow and seedlings are identical.

Any system of classification must obviously place caraway close to the carrot. Their seedling appearance, patterns of growth, and flowering are obviously similar. The caraway plant will be a lower, stockier seedling, its leaves not as finely articulated as those of the carrot. Caraway's next closest non-weed look-alikes are dill and fennel. On closer observation these latter seedlings have round stems and leaves that are more threadlike and lacy.

Caraway should appear in about two weeks. Caraway and fennel do not grow well together, but caraway and peas reportedly aid each other's growth.

Caraway

Coriander

Umbelliferae
  Coriandrum sativum

The origins of the coriander plant have been obscured by time. What is certain is that it is one of the most ancient of cultivated herb plants. The Chinese used both roots and seeds before 5000 B.C.; the earliest Greek, Roman, and Egyptian civilizations also knew it well. Coriander had appeared in England before the Norman Conquest, and this annual herb was one of the first arrivals in the New World.

Coriander is best sown directly where it is to grow, as it does not transplant well. There are many similarities between this seedling and those of the plain leaf or Italian varieties of parsley. Coriander is sometimes called Chinese parsley. The cotyledons of both plants are identical at first, but those of the coriander then continue to grow, becoming longer and narrow. The next appearing leaves also offer confusion. Coriander's leaves are a darker green, a bit flatter, and a bit more squared off and sharply serrated than the longer-stemmed, broader-leaved parsley. When older, some varieties of coriander show reddish stems.

If you are still in doubt, one very accurate method of identification is to taste a leaf. Be forewarned that coriander is the major ingredient in curry powder.

Coriander seedlings will appear after about two weeks of favorable weather. This plant is quite apt to reseed itself, and its seeds will over-winter. Coriander hinders seed formation in fennel, so plant these two some distance from each other.

Coriander

Plain Leaf Parsley

Curled Leaf Parsley

Umbelliferae
  Petroselinum hortense

Umbelliferae
  Petroselinum crispum

The herb parsley is a close relative of celery. Like celery, parsley is native to the Mediterranean region, though it spread rapidly to many other lands by the first century B.C. Both leaf types, plain leaf and curled, were described in the fourth century. Like celery and its thick-rooted alternate celeriac, parsley also takes two forms. Both turnip-rooted or Hamburg parsley and the varieties grown for their green leaves were in use long before the celerys, however.

The parsley seedling makes its first appearance with a pair of spear-shaped cotyledons. This is a familiar shape among our garden herbs and vegetables; the seed leaves of the related celery, parsnip, and coriander all are spear-shaped at first, as are those of the unrelated tomato, pepper, and eggplant.

After the parsley seedlings have grown a few leaves they are most apt to be confused with the seedlings of celery, celeriac, and coriander. Curled or moss-curled cultivars of parsley distinguish themselves quite readily by their overlapping, finely scalloped leaves. The plain leaf or Italian parsley has flatter, simpler leaves that are somewhat larger than those of the slow-growing celery/celeriac seedlings.

The first leaflets of the coriander seedling soon lengthen, losing their initial resemblance to the parsley cotyledons. The leaves that follow are flat, like those of the Italian parsley, but more crisply divided.

Parsley seeds are slow to germinate, usually requiring at least two weeks to appear. Parsley is a biennial herb but is usually grown as an annual because it produces the most foliage in its first year, sending up a flower stalk in its second spring. It will reseed itself if not harvested before flowering, and its seeds are winter-hardy.

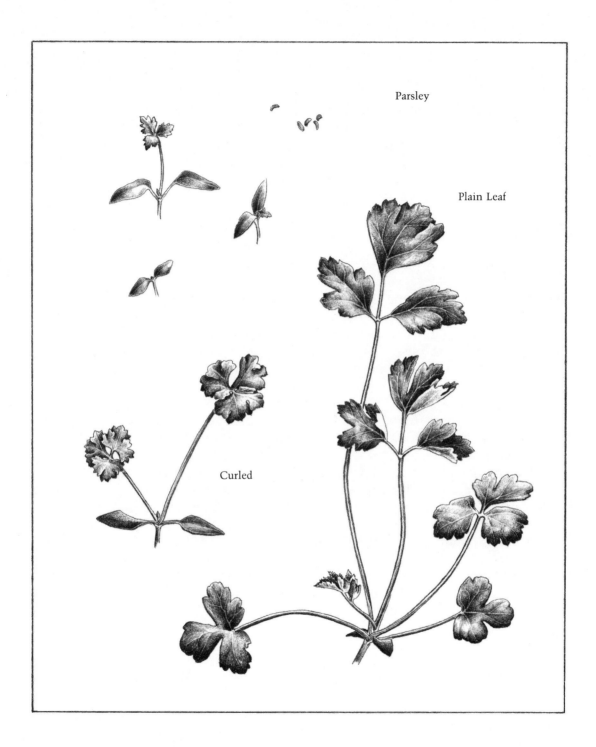

Parsley

Plain Leaf

Curled

Sweet Potato

Convolvulaceae
  Ipomoea batatas

By some undocumented method, perhaps by very early Spanish explorers, the sweet potato found its way from its South American homeland to the Philippines and the East Indies. Portuguese travelers then carried it to India, China, and Malaya. Because of its very early and unobtrusive introduction to the Pacific and Far East, biologists cataloguing the origins of plants at first surmised that the sweet potato had evolved in Asia.

The sweet potato was introduced to Virginia around 1648. Because it is a warmth-loving plant, it has been a much more important crop in the southern states than in the north. In the south it is used extensively both for human consumption and as stock feed.

The sweet potato, like the white or Irish potato, is not normally propagated by seed. This vine cannot be convinced to flower or set seed except in tropical climates. Sweet potatoes are started from "slips," or offshoots of a sprouting mature tuber. These offshoots have two or three leaves and the beginnings of a root system when they are broken from the parent tuber and transplanted in the field.

Sweet potatoes come in many colors and two basic textures; dry-fleshed or moist-fleshed. The moist-fleshed types are frequently and incorrectly called "yams." True yams are of the genus *Dioscorea* and are rarely seen outside the tropics.

Once one has gone to the considerable effort of raising and transplanting slips or purchasing and planting the pre-started shoots, one rarely misplaces them. The sweet potato is of the same family as the ornamental morning glory and the dreaded bindweed. The resemblance is easy to see, especially in these young slips.

The sweet potato has smooth gray-green leaves and quickly acquires its vining habit. The stems are sometimes a blotchy red and green. The shape of the leaves is initially very simple—in outline rather like the first bean leaves. Sweet potato leaves branch alternately from the stem, and their shape becomes a bit more complicated as the plant grows older.

Sweet Potato

Sweet Basil

Labiatae
  Ocimum basilicum

Sweet basil is one of the most popular seasonings in the world and certainly a favorite cooking herb in the United States. It is native to western and tropical Asia, and was singled out of the wild to be pampered in the gardens of the most ancient civilizations. Fifty or sixty species of basil are now grown; the most familiar in this country is the sweet basil, *Ocimum basilicum.*

Basil is easily identified as a seedling primarily because, compared to most of our common herbs, it is large and fast-growing. It begins small with the same pair of tiny oval cotyledons possessed by innumerable unintended weeds and many herb relatives. These cotyledons rapidly grow larger into broad spade-shaped leaves. The next leaves always appear in pairs opposite each other on a decidedly square-sided stem. These leaves are moderately serrated on the edges and buckle and bend over backwards in an exaggerated fashion. They are very tender, hairless leaves, and tear easily. The strong basil odor is present even without bruising a leaf.

  "Dark Opal" varieties of sweet basil are also gaining in popularity. These plants are identical in form to the green basil but have an ornamental purple coloring throughout. They cannot easily be confused with any of our common herbs or weeds.

Basil seedlings appear readily in about five days. They are very susceptible to frost damage.

Sweet Basil

Sweet Marjoram   Oregano

Labiatae
  Origanum majorana (Majorana hortensis)

Labiatae
  Origanum vulgare
  O. heracleoticum

Marjoram and oregano have many characteristics in common—both structural similarities and flavor/odor similarities. This group of herbs has been claimed at one time or another in past centuries to be a medicinal cure for virtually every human ailment and dissatisfaction. There is still some confusion over the most accurate naming of these plants. The plant commonly known as sweet marjoram is native to Portugal and other Mediterranean climates. Pot marjoram, *O. vulgare* (sometimes known as wild marjoram), also hails from the Mediterranean region, as does *O. heracleoticum*, or winter marjoram. Both pot marjoram and winter marjoram are also known as oregano.

Sweet marjoram is a tender perennial which reaches a height of about twelve inches and has white or pink flowers. Pot marjoram is a taller, hardier plant with flowers varying in color from white to violet; most of its uses are medicinal. Winter marjoram is better tasting than pot marjoram and is a lower growing plant with white flowers.

As one might guess, all of these species of the genus *Origanum* are identical as seedlings. Because the seeds are almost dustlike in size and the resulting seedlings so small, marjoram and oregano are usually started indoors in order better to observe their progress and insure that weed competition is not overpowering. These are slow-growing, trailing seedlings with oval, short-stemmed leaves. The leaves are dark green and cupped backwards, and emerge in pairs on the reddish stems. Additional shoots soon appear in the lower leaf axils.

Sweet marjoram seedlings appear in eight to ten days. Pot marjoram and winter marjoram, or oregano, take a little longer, about fourteen days.

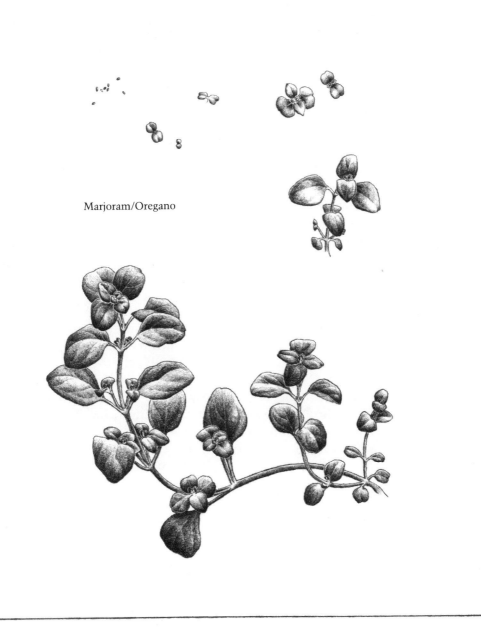

Marjoram/Oregano

Spearmint

Peppermint

Labiatae
  Mentha spicata

Labiatae
  Mentha piperita

According to most sources, all of the many species and varieties of mint were originally natives of the Near East. They spread rapidly, with and without human intervention, throughout the world. They were clearly well-known to the authors of the earliest herbals of diverse nations, as well as to the recorders of Greek and Roman mythology and the Christian Bible.

Most of our common mints are perennials that spread on underground runners. They are naturalized as weeds in North America as well as on most other continents. All of the mint varieties will readily hybridize, causing much confusion as to their proper identification.

Spearmint and peppermint are the best known of the mints. The lavender-flowered peppermint is now most often used for flavoring medicines and candies. Peppermint plants will readily form viable seed, but because the plant is a hybrid, the seed cannot be counted upon to produce plants that will imitate the parent. The mature spearmint plant shows a more compact growth and its flowers can vary in color from off-white to purple; this species has a stronger flavor and is more apt to be used as a flavoring in drinks and foods. This is the variety of mint most frequently offered in seed catalogues.

Spearmint and peppermint are difficult to differentiate in the early stages of growth. From their tiny two-leafed beginnings, mints quickly grow to upright seedlings. As with the other members of this family, the leaves appear in pairs opposite each other on a faintly square-sided, purplish stem. The leaves are pointed, slightly serrated on the edges, and give off a mint odor that is easily identifiable to anyone who has ever chewed gum. The mints waste no time in sending up additional sprouts from extensions of their roots.

English pennyroyal seedlings, *M. pulegium*, which are also mints, resemble the illustration opposite to some extent. This latter plant, however, remains a very low-growing, creeping herb with tiny round leaves. Apple mint, *M. rotundifolia*, is yet another common mint which can be found growing wild. Its leaves are gray-green, round, and fuzzy, as another one of its common names, woolly mint, suggests.

Mint seedlings should appear in about ten days. The seed does not remain viable for more than a year.

Mint

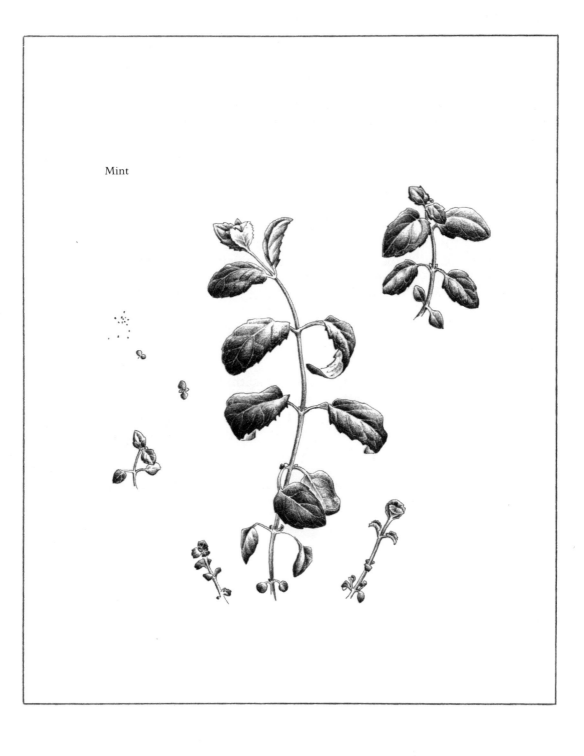

Rosemary

Labiatae
   Rosmarinus officinalis

Rosemary grew originally in the western Mediterranean, perhaps along the coasts of France and Spain. In its homeland it is a perennial evergreen shrub reaching a height of four feet or more. Roman conquerors may have brought rosemary to northern Europe and England, or perhaps the returning Crusaders should be credited with advertising its virtues. However it spread outward from its homeland, it was quickly invested with all sorts of powers and ritual uses, and cultivated for its flavoring in the kitchen.

For many of us in the northern United States, rosemary can be a difficult herb to cultivate, as it is sensitive to frost and must be protected during the winter months. Rosemary is most often propagated by cuttings or layering rather than by seed because only ten percent of its seeds can be expected to germinate. In the northern states rooted plants are not always available and so the tiny seeds must be sown.

Rosemary is a dicotyledonous plant which shows its insignificant pair of seed leaves above the soil. Soon pairs of narrow, pointed leaves begin to appear. They are dark green in color, lighter gray-green beneath, with indented veins giving the surface a bumpy appearance rather like that of the downy sage leaves. Rosemary is a small and slow-growing seedling. It is susceptible to accidental weeding if care is not taken to separate it from similar-looking but unwelcome weed seedlings.

Sow rosemary seed very thickly and wait patiently for at least twelve days, often longer, for a few seedlings to develop.

Rosemary

Sage

Labiatae
  Salvia officinalis

Sage is yet another Mediterranean native. There are many species and varieties of sage, different in coloration and leaf shape, but *S. officinalis,* the one depicted here, is by far the most commonly grown in North America. The genus name, *Salvia,* means "health" or "salvation," indicating its early and important use as a medicinal herb of broad applications.

This perennial herb is easily grown from seed and will obligingly reseed itself if left to flower. The seedling begins with two dicotyledons similar in size and shape to those of the other plants of this family. Like its cousins, sage sports its leaves in pairs opposite each other on squarish stems. These stems, as well as the long oblong leaves, are covered with a downy fuzz. The stems are white at first, but may later turn quite purple as the plant matures. The leaves are distinctive; their veins are indented, giving the gray-green surface a soft, bumpy appearance. Sage seedlings, and even mature plants, can be confused with the herb horehound, to which they are related. Horehound also has fuzzy leaves with a similar puckered surface, although the leaves are rounder in shape with scalloped edges. Woolly mint, or apple mint, also resembles sage, at least until its strong mint odor is noticed.

These seeds, large by herb standards, should develop into visible seedlings in about twelve days.

Sage

Thyme

Labiatae
  Thymus vulgaris

Thyme is one of the basic components of the contemporary herb garden, as it was in gardens more than two thousand years ago. Native to the coasts of the Mediterranean, thyme is a hardy perennial herb and has been naturalized in all but the most extreme climates. There are several closely related species of thyme. The drawing opposite is of the most popular variety, French thyme. All of the related species and varieties will resemble one another as seedlings, differing later in stature, color of flower, and size of leaf.

Thyme is often and easily started from seed, although it can also be propagated by root divisions or cuttings. Thyme is an exceptionally small seedling even for an herb. If sown directly out of doors this viney sprig will find tough competition from the weeds and weather. A carelessly weeded thyme plant quickly announces the mistake by its strong but pleasant odor. Thyme has opposite branching leaves on a stem which is not quite round. The leaves are triangularly pointed and will always be very small, even on the mature woody plant.

Thyme seedlings should appear in about ten days.

Thyme

Tomato

Solanaceae
  Lycopersicon lycopersicum esculentum

The tomato is for most home gardeners the most important member of the family *Solanaceae*. Originating in and around the Andes region of South America where it grows as a perennial vine, the tomato in its many shapes and colors has been cultivated there for thousands of years. The earliest explorers in South America brought this strange-looking vegetable home with them as a curiosity. Tomatoes quickly became popular in the countries of the Mediterranean region, but were not accepted as an important food plant in Europe or North America until much later. Tomato culture finally caught on in this country a little more than a hundred years ago. Modern plant breeding has improved the disease-resistance and the size and smoothness of the fruit. "Beefsteak" type tomatoes are a recent development. Science has also developed uniform, long-keeping varieties with less nutritional value and almost no flavor.

Tomato seedlings initially resemble pepper and eggplant seedlings, and to a lesser extent, Swiss chard and beet. Soon tomatoes become easily identifiable by their fuzzy stems and leaves, oily feel, and strong aroma. There *is* some slight hope for those who wish to know whether an unexpected volunteer seedling or an illegible or unlabeled windowsill plant might produce regular-size or smaller fruit. The cherry-type tomatoes will most often have leaves that are more complicated and deeply divided than those of the standard varieties, and some of the other smaller-fruited types will have leaves with a downy quality due to the fuzziness of the leaves and slightly indented veins. The seedling differences between varieties are minute—correct identification depends on a vague impression rather than an easily defined or illustrated trait.

Tomato seeds are very hardy, and stray fruits from a previous planting are quite apt to disperse their seeds to other regions of the garden. When contemplating the fate of unplanned spring seedlings, remember that if last year's crop were hybrid varieties, the offspring might not produce as well as the parent plant did.

Tomatoes are warmth-loving plants and will appear in about eight days at a soil temperature of 85 degrees F.

Tomato

Pepper

Solanaceae
  Capsicum frutescens

The wild forms of both sweet and hot peppers are found in the same tropical South American area as the ancestors of the tomato. Peppercorns and ground pepper are from the berries of an entirely different plant, *Piper nigrum*. The expression "pepper" as used for our familiar garden peppers is probably another of Christopher Columbus's irretrievable misnamings. He and his crew were the first Europeans to sample the South American pepper; they were thrilled to find a hot spice to rival the peppercorns from the Caucasus region, a valuable trade item. They managed to overlook several other plants that would later prove to be far more important to civilization. The potato, for instance.

The spear-shaped cotyledons of the pepper have much in common with tomato and eggplant seedlings as well as many of our herbs and weeds. Pepper seedlings are easily distinguished from the related eggplant, tomato, and potato seedlings by their smooth, shiny leaves. The pepper's stem is upright and woody.

Unfortunately, it is impossible to differentiate between the sweet peppers and the common varieties of hot-flavored peppers at these early stages. The drawing opposite happens to be of a green bell pepper but it could just as well be a cayenne seedling.

Like tomato and eggplant seeds, pepper seeds prefer warmth for rapid germination and growth. Seedlings should appear in ten days under favorable conditions.

Pepper

Eggplant

Solanaceae
  Solanum melongena

The eggplant is similar in flower, structure, and habit to the South American vegetables tomato, pepper, and potato, and is therefore included in the same botanical family. The eggplant originated in far different lands, though; it was cultivated in and around northern India long before written language could record the event. Apparently this species was introduced to Europe by Arabic peoples during the Dark Ages, but was not well known there until the middle of the sixteenth century. Even then it was grown merely as an ornamental and not until somewhat later as a food plant.

Distinctly different types of eggplant were also developed in ancient China. The oriental varieties are generally more prolific, have smaller fruit, and a slightly different flavor from the eggplants of Indian origin with which we are more familiar. We normally think of eggplant fruit as being a shiny, dark purple, but variegated, yellow, brown, white, pear-shaped, long-fruited, and even green eggplants have been grown by various peoples at various times.

All varieties of eggplant start out life looking much like tomato or pepper seedlings, and in many areas of the country they are started indoors at about the same time as these other vegetables; they too need a long, warm growing season. Soon the eggplant's soft, rather shapeless, friendly leaves appear, dispelling any confusion. Some varieties of eggplant have striking dark purple stems and veins, and many develop sharp bristles or thorns on the stems and undersides of their leaves.

Eggplant seedlings are slow to appear. They will emerge in about fourteen days if a soil temperature of 75–90 degrees F. is provided.

Eggplant

Potato

Solanaceae
Solanum tuberosum

Congratulations. You now belong to a very small group of individuals who know what a potato seedling looks like.

The reason for all this mystery is that almost no one plants potato from seed. The traditional and quicker method is to plant "sets" or "eyes" of existing tubers. Most potatoes are highly hybridized, which means that plants resulting from its seed are not of a uniform quality. Recently, however, a few seed catalogues have begun to offer potato seed for sale under the cultivar name of "Explorer" that will give consistent and satisfactory results.

Planting from seed eliminates the problems encountered with seed tubers, which require delicate handling and storage until the next growing season, and also removes the threat of tuber-borne viruses and diseases. However, for the purposes of the backyard gardener, the extra effort involved in growing potato seedlings and the longer period needed for the tubers to reach maturity makes planting from seed not particularly worthwhile except as an interesting experiment.

Potatoes are another plant with beginnings in South America, most likely in the Peruvian Andes. The tubers of *S. tuberosum* and its relatives have been used as a major food source for the peoples there since time immemorial. Europe didn't discover the potato until about 1537, when Spain invaded what is now Colombia. The Spaniards brought it home with them and its use then spread rapidly throughout the European continent.

Potato seedlings are not easily confused with their vegetable relatives in the family *Solanaceae,* although they do tend to resemble miscellaneous weed plants simply because they are small and have no particularly striking features. From its very diminutive and fuzzy-stemmed beginnings, the potato slowly grows into a viney sprig with heart-shaped, backward-cupped leaves.

This seedling bears little resemblance to the clusters of large dark-green, deeply-veined leaves that emerge from the planted seed tuber. The older potato leaves do look like tomato leaves, and in fact tomato and potato plants are so similar physically that it is possible, though not particularly useful, to graft the shoot of a tomato plant to the stem of a potato.

It takes one and half to two weeks for potato sprouts to emerge from seed. Plantings from sets are much quicker, usually appearing within a week.

Potato

from seed

from set

Pumpkin

Cucurbitaceae
  Cucurbita pepo

To a botanist, the word "pumpkin" may denote a plant that has, among other characteristics, fruit with a rind that never becomes very tough, even when mature, and hard, furrowed stems. This definition includes many plants whose fruit we commonly call squash. For the purposes of this book, however, the commonly accepted usage of the word "pumpkin"—a plant with yellow-orange, round fruit—is intended.

The species *C. pepo* is probably native to Mexico and Central America. This group encompasses many cultivars, including most of our common pumpkins, zucchini, yellow summer squash, English marrow, scallop, golden custard, acorn squash, and some small gourds. Unfortunately, the variations in the cultivar names are due only to differences in the harvested product and do not necessarily indicate a change in leaf shape.

The specific cultivar illustrated on the facing page is a "Small Sugar" pumpkin. It is drawn to a scale approximately one-third life-size. This entire family of vegetables makes its first appearance in the garden with nearly identical pairs of broad, oblong cotyledons. The first pumpkin leaf to open fully is basically round with sawtooth edges, as are the next several leaves to follow. The fourth and fifth leaves to unfurl begin to hint at the deeply lobed shape of the leaves of the mature plant. The stems and leaves of all varieties of *C. pepo* are harshly bristled and musty-smelling.

It is obvious from the older seedling shown that this variety is a vine, as are most pumpkins, and has an already lengthening stem. The leaves are spaced much farther apart than on bush varieties of the *Cucurbita* genus. This characteristic immediately separates the vining pumpkin from some of the confusing related varieties that take bush forms. Most problems of identification will occur with vining scallop-type and marrow squashes, vegetable spaghetti, and some of the small ornamental gourds.

All of the many cultivars of *C. pepo* will readily cross-pollinate with each other, and rare crosses may even occur with other species of this genus. Winter-hardy seeds left to sprout next spring should be regarded with suspicion, however, as the fruit of these crossings may not be worth the time spent tending the plants.

Pumpkin seedlings should appear in about eight days.

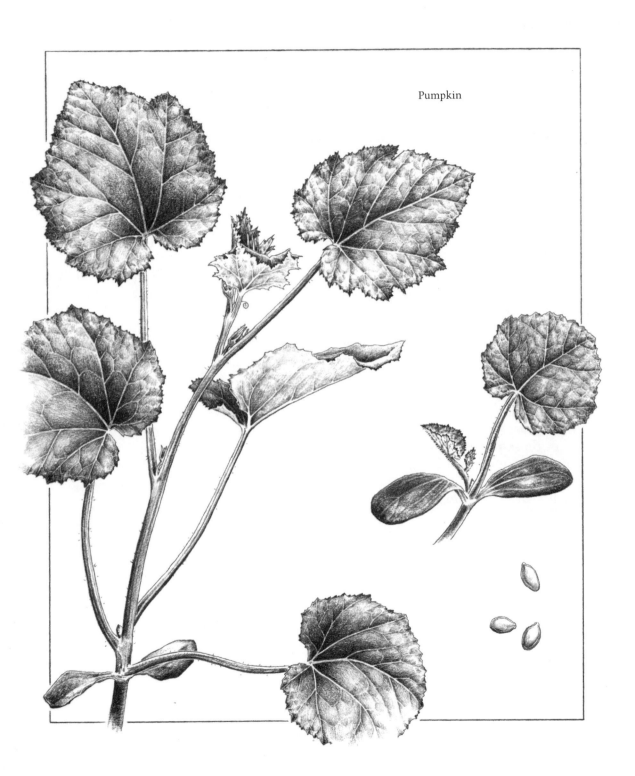

Pumpkin

Zucchini Squash

Cucurbitaceae
  Cucurbita pepo

One of our most common squashes, this plant is technically considered to be a pumpkin by some botanists, which explains its resemblance to the orange-fruited plants we commonly know by that name. The zucchini or cocozelle is often associated with Italian cuisine, but originally this plant was known only to the natives of North and Central America and Mexico, who were growing similar varieties even before the first century A.D.

The illustration of the zucchini seedlings is approximately one-third the normal size of the plants. Zucchini seedlings announce their arrival with a large pair of flat, oval cotyledons; these are identical to most other initial sprouts of this family. The first several leaves that open thereafter are basically round, with sharply serrated edges.

There are sufficient similarities between the various cultivars of this species to provide much confusion amongst the poor garden-note-takers among us. All will have harsh, bristly, hollow stems and leaves with tiny spines; they are quite unpleasant to handle.

As the drawing of the larger seedling shows, by the time the zucchini sends forth its fifth leaf we can really begin to see the deeply lobed character of the more mature leaves. These "fingers" of the zucchini leaf will be more pronounced than those on any of our other common bush squashes (for instance the yellow summer squashes) of similar age. Zucchini is usually grown in bush form; the stem distance between the leaves is much shorter than it is on the vining squash plants and pumpkins.

Contrary to popular belief, a zucchini will not cross-pollinate with a cucumber or any type of melon, these latter being of a different genus. They are cross-fertile with all of the pumpkins, squashes, and gourds of this species, and on rare occasions they will perversely cross-breed with other species of *Cucurbita*, for example a butternut of the species *C. moschata*. This mixed breeding will only affect any volunteer seedlings that appear the next year, and not the current season's produce.

Seedlings should appear in about seven days.

Zucchini Squash

Yellow Summer Squash

Cucurbitaceae
  Cucurbita pepo

This is yet another vegetable that originated in the warm regions of the American continent, where it and other squashes and pumpkins played a major role in diet and agriculture for millennia. This species encompasses a remarkable diversity of shapes, colors, and sizes of fruit due to its remarkable propensity for cross-breeding.

The illustration on the page opposite is of a popular cultivar of yellow crookneck squash. The scale is approximately one third life-size. There are many shapes of yellow summer squash, but the plants themselves are all virtually identical as seedlings. Most are bush varieties, making it possible to distinguish them easily from their close relatives that are known to be vines. The leaf stems of bush varieties are situated quite close together on the central stem as compared to those of the vining plants.

Vine or bush, plants of this family all appear with a pair of dark, rather shiny, oval cotyledons that are unmistakable because of their large size. The first couple of leaves to open out will be identical to most other cultivars of this species. The foliage and stems are also similar, harsh to the touch with bristles and small spines on stem and underside of leaf. Most confusion occurs with other bush varieties such as scallop and custard squash and some varieties of small gourds. The leaves of the very popular zucchini cultivars, also members of this species, are much more deeply lobed and serrated than those of the yellow squash at a similar age. Most varieties of acorn squash will have rounded leaf lobes.

The other common species of this genus, *C. moschata*, *C. maxima*, and *C. mixta*, described in the following pages, are a bit more pleasant to the touch and their leaves do not become as deeply divided as those of the *C. pepo*.

Squash and pumpkin seeds will over-winter and sprout in spring if fruit is left in the garden. Note the warnings about cross-pollination given under the descriptions of zucchini and pumpkin.

Like others of this species, yellow squash seedlings should appear seven days after sowing.

Yellow Summer Squash

Acorn Squash

Cucurbitaceae
  Cucurbita pepo

All of the pumpkins and squashes grown world-wide were originally cultivated in South and Central America, Mexico, and southwestern parts of North America. They were (and are) of major importance to the agricultural civilizations that grew them in a wide variety of shapes and sizes even before the time of Christ.

The harvested acorn squash looks so different from, for instance, a pumpkin or zucchini, that it is hard to believe they are considered by botanists to be merely different cultivars of the same species. The illustration shows two stages in the life of a "Table Queen" acorn squash. The drawing is approximately half life-size. The two flat, oval cotyledons appear first, followed by a single, round, prickly leaf. This first leaf, and the next ones, are disturbingly similar to the related squashes and pumpkins of the same species.

One way to separate their identities, if you are included in that large group of poor garden-note-takers and some confusion has occurred, is to note which cultivars planted are bush-types and which are vines or semi-bushes. Acorn squash is usually a vine or semi-bush; this means that it will have quite a long distance on the main stem between the leaf stalks compared to, for instance, the zucchini, which is almost always a bush cultivar.

Most of the acorn squashes have leaves that are not as deeply divided, and lobes that are not as acutely pointed as those of the other commonly grown cultivars of this species.

See the notes on zucchini and pumpkin for information about potential cross-pollination and its effect on seeds that may over-winter in the garden.

Seedlings should be sown after the ground has warmed, and will appear in about nine days.

Acorn Squash

Butternut Squash

Cucurbitaceae
 Cucurbita moschata

The butternut is one of our most popular winter-type squashes. This plant shares the species name with other types of squash and pumpkin, for example "Sugar Marvel," "Turkish Honey," "Kentucky Field," "Dickinson," and "Golden Winter Crookneck." All varieties of *C. moschata* are native to the North American southwest, Mexico, and Central America. The cultivation of this and similar varieties was widespread by the time of European contact in the fifteenth century.

The illustration on the opposite page is drawn approximately half life-size and shows a "Waltham" butternut with one leaf opened as well as a later drawing of the plant with four leaves unfolded. Butternut's first appearance is with the oval, flat "seed leaves" common to all of the other plants included in this family.

It is obvious from the first leaf or two that this butternut and the other squashes and pumpkins of this species have much softer, simpler leaves than do the related vegetables described on the preceding pages. Their stems are quite hairy but without the coarse bristles and spines. Their leaves are not as deeply lobed or serrated on the edges.

In the illustration of the older seedling, the fourth leaf to unfold, counting up from the base, shows clearly the pattern that the later leaves will follow.

Most problems of identification will occur with cultivars of *C. maxima*, such as the hubbards and buttercup squash, and with the less common *C. mixta* varieties such as cushaw and Japanese pie pumpkins. The leaves of these species are somewhat more deeply lobed than those of the *C. moschata* cultivars.

The different varieties of this species will cross-pollinate with each other. Cross-fertilization with the plants of *C. pepo* and *C. maxima* is also possible. The quality of the fruit of the seedlings that reappear voluntarily from seeds left over winter in the garden may not justify the effort of tending them.

Seedlings of winter squash usually take nine days to emerge.

Butternut Squash

Hubbard Squash

Cucurbitaceae
Cucurbita maxima

The hubbard squash is a popular winter squash and therefore a good representative of the species *C. maxima.* Some of the other cultivars of this species are "Mammoth" and "Mammoth Chili," "Delicious," "Turk's Turban," "Boston Marrow," "Essex Hybrid," "Marblehead," and banana and buttercup varieties. This type of squash was found only in Brazil, Bolivia, Chile, and Peru at the time of European exploration. It has now, of course, spread to almost all areas of the globe.

The hubbard seedling depicted on the page opposite is approximately one-third its natural size. It began life above ground with a pair of enormous cotyledons, larger than those of most other squashes and pumpkins, obviously larger than those of the cantaloupes, watermelons, and cucumbers. Its first leaves are quite simple in outline, as are the leaves of the mature plant. This immediately distinguishes the cultivars of the species *C. maxima* from all squashes and pumpkins of the closely related species *C. pepo,* mentioned earlier.

Many mix-ups occur in the identification of the hubbards and buttercup squashes of this species and the equally popular butternut of the species *C. moschata.* All of these are usually vine crops and in addition they all have comparatively simple leaf shapes. On close examination, however, the butternut and its relations will be found to have softer, hairy leaves and vines, and only slightly lobed and finely serrated leaves. The hubbard, for example, will have larger, wavy-edged leaves that are more complicated in outline.

The less common vegetables of the *C. mixta* species, which includes cushaw squash and Japanese pie pumpkins, may also resemble plants of *C. maxima* and *C. moschata.* The cultivars of *C. mixta* are always vines, are intolerant of cool weather, and have fairly simple leaves with soft hairy stems.

The members of *C. maxima* are somewhat cross-fertile with other species of squash and pumpkin as well as with cultivars of their own species. Consult the warnings given under pumpkins and zucchini if considering the fate of seedlings that appear from hardy over-wintered seeds.

Most winter squash seedlings emerge about nine days after sowing.

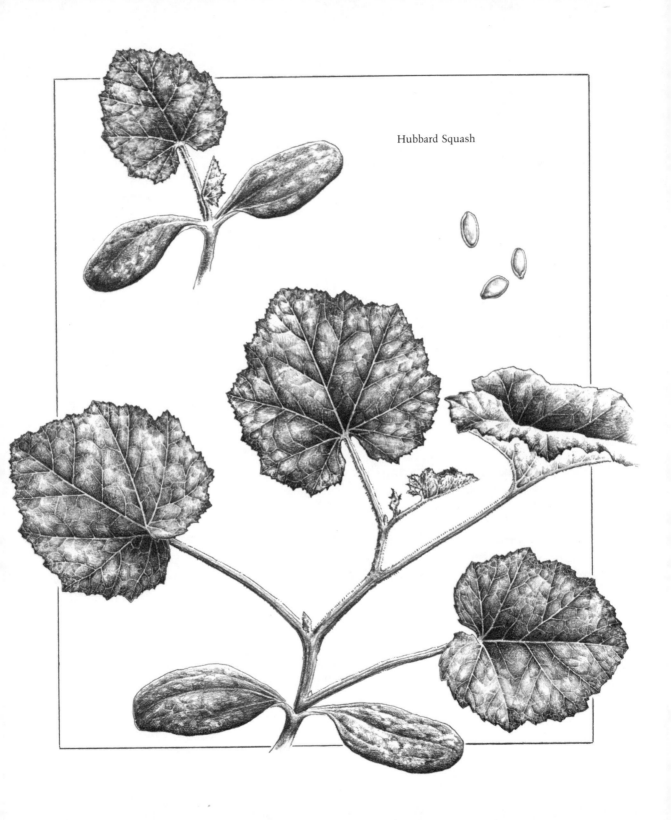

Hubbard Squash

Cucumber

Cucurbitaceae
  Cucumis sativus

Cucumbers have been cultivated for more than four thousand years. The cucumber is believed to be native to northern India, and was carried westward long before written history could record its travels. By the second century B.C. it had found its way to China, and its popularity had spread extensively by the beginning of the Christian era. Cucumber cultivation had reached Europe by the ninth century A.D.

Columbus is credited with introducing the cucumber to the New World; it was planted on Haiti and adjacent islands in 1494. Native Americans in touch with early Spanish explorers quickly adopted the cucumber, and Europeans who reached Virginia in 1584 mentioned finding cucumbers already growing there when they arrived.

The illustration of the cucumber and its seeds was drawn about one-fifth smaller than life-size. All cultivars of this species will be identical at the seedling stage. The West Indian gherkin, *Cucumis anguria,* is the true gherkin used for pickling. Its mature fruit is round and warty, about one inch in diameter; the leaves and vines of this related cucumber are smaller than the common pickling and slicing cucumber illustrated.

The cucumber seedling is one of the few of this most confusing family of plants that can quite easily be distinguished from its seedling relations. The cucumber has smaller cotyledons than the squashes and pumpkins—they're more similar in size to the cantaloupe's. This seedling will continue to look much like the cantaloupe seedling until several leaves have completely unfolded. All of the common varieties of cantaloupe will have almost hairless leaves that are quite shiny, serrated on the edges but simpler in outline than the cucumber's leaves, which are a bit bristlier as well. The three-cornered shape of the cucumber leaf becomes more pronounced in later leaves, and thus more obviously different from the cantaloupe leaf with its less dramatic shape.

Contrary to myth, cucumbers will not cross-pollinate with any of the other *Cucurbitaceae* members. Cucumber seedlings usually appear about seven days from planting. Seeds will often overwinter if fruit is left in the garden.

Cucumber

Muskmelon

Winter Melon

Cucurbitaceae
  Cucumis melo var. *reticulatus*

Cucurbitaceae
  Cucumis melo var. *inodorous*

There are two botanical varieties of *Cucumis melo* that are important in the backyard gardens of this country. Muskmelon or nutmeg melon is a salmon- or green-fleshed melon with a textured or netted rind. Also grown extensively here are the smooth-skinned honeydew or winter melons which include the casaba and crenshaw types. All of these melons are frequently called cantaloupe, although this name more accurately refers to a variety, *C. melo* var. *cantalupensis*, raised in Europe and rarely encountered in North America.

All of the members of this genus are believed to be native to western Asia and India, although much is unknown about their early development and their spread to other lands. No specimens resembling these melons have been found growing wild. They have been cultivated for at least four thousand years.

The illustration is of a "Far North" muskmelon and is only slightly smaller than life-size. Muskmelon and winter melons are identical as seedlings. Both varieties of melon appear above ground with a pair of flat, oval cotyledons. These are basically similar in form to the cotyledons of all other members of this family, and are the same small size as those of the cucumbers.

Most cultivars of melons are vines, as indicated by a rapidly lengthening main stem, with leaves spaced far apart compared to those found on bush varieties of this family. Cantaloupe vines are smaller in leaf and stem than squash or pumpkin vines, and much less bristly.

Most confusion will occur with cucumber seedlings, especially when only the first leaf is showing. Cucumbers are also usually thin vines, but their leaves are more sharply three-cornered, with more acutely serrated edges. Cantaloupe leaves are quite smooth, unlike the hairy cucumber leaves, and are faintly bluish. The edges of mature melon leaves are wavy and do not lie flat.

All of the botanical varieties and cultivars of *Cucumis melo* will cross-pollinate with the help of insects. This will not affect the first harvest, but if seeds are left in the garden to sprout in the spring, the crop from these second-generation plants may show the results of cross-breeding.

Cantaloupe seedlings should appear in about a week if sown in warm soil.

Cantaloupe

Watermelon

Cucurbitaceae
  Citrullus vulgaris

The cultivation of watermelons began in prehistoric times in Central Africa. Early on their use spread unobtrusively and by unknown routes to Egypt, the countries of the Mediterranean, and India, where they became so commonplace that it was at first presumed that watermelons were of Asian descent. David Livingstone, famed explorer and missionary to Africa, discovered watermelons growing in a truly isolated spot there in the mid-1800s, thus confirming their African origins.

Today the watermelon is still sometimes cultivated in the semi-desert areas of its homeland as a source of water during dry seasons. Watermelons arrived in this country with the earliest European colonists.

The two large, flat cotyledons of the watermelon unaccommodatingly look a great deal like those of other melons and squashes. The leaves that follow, however, are unmistakably different. The smooth, somewhat shiny leaves have an unusual blue-green color not found in any of the other vines likely to appear in our gardens.

More obvious than their subtle coloring is the distinctive shape of the leaves, which is rather like that of an oak leaf. Each successive leaf then tries to outdo the last in further complexity of edge; the oak leaf analogy is obsolete by the fourth leaf. These leaves are generally smaller than the plain green leaves of the related melons, cucumbers, squashes, and pumpkins. Watermelon vines have none of the strong, musty odor of most of their relations, although they do have similar bristly stems.

Watermelon seedlings are fairly sensitive to cold weather. They should appear eight days after sowing in warm soil.

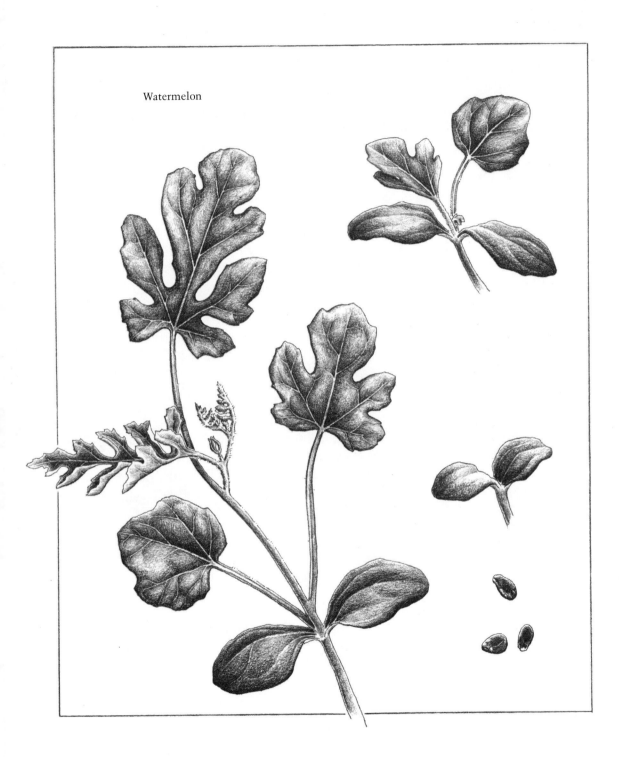

Watermelon

Lettuce

Compositae
  Lactuca sativa

The wild form of lettuce, *L. scariola,* from which our garden lettuces are derived, was first gathered as food in Asia Minor and central Europe, and its popularity spread quickly to other regions. The Greeks wrote of its cultivation in the fifth century B.C. There are now four different botanical varieties of cultivated lettuce. *L. sativa* var. *crispa* encompasses all of the loose-leaf type lettuces, which most closely resemble the earliest lettuce forms. *L. sativa* var. *capitata,* or the hard-heading lettuces, are a comparatively recent development dating from the Middle Ages. Cos or "Romaine" lettuce, *L. sativa* var. *longifolia,* was first grown near Italy, as its common name implies. Less well known is *L. sativa* var. *asparagina,* stem lettuce, also known as celtuce. This last variety has thick stems which are eaten like celery.

Lettuce is one of the relatively few vegetables of which it is possible to discern differences between cultivars at an early stage, but only if those differences involve coloring or the shape of the leaves. This knowledge was useful last spring when I planted a row of lettuce, one half a loose-leaf type and the other half "Romaine." One end germinated poorly, and although I once again forgot to write down which end was which, I could easily tell which seed to replant by recognizing the type that had germinated successfully.

All lettuce looks pretty much the same until the fourth or fifth leaves have formed, the first leaves appearing woefully similar to all the tiny round weed leaves coming up everywhere. Some lettuce cultivars will look like endive and escarole as seedlings. The differences are minute—the endives have slightly tougher leaves, a bit hairy on the underside.

The illustration shows four of the most commonly grown cultivars of lettuce. "Iceberg" has quite jagged leaves compared to the loose-leaf lettuces, represented by "Oak Leaf" and "Black-Seeded Simpson." The leaves of the "Romaine" are a little less tender and show a more upright growth, even as young seedlings.

Lettuce is a cool-weather crop and germinates within six to eight days at a soil temperature of 60–65 F. Cover the small seed thinly with light soil, as its germination is further assisted by a small amount of light.

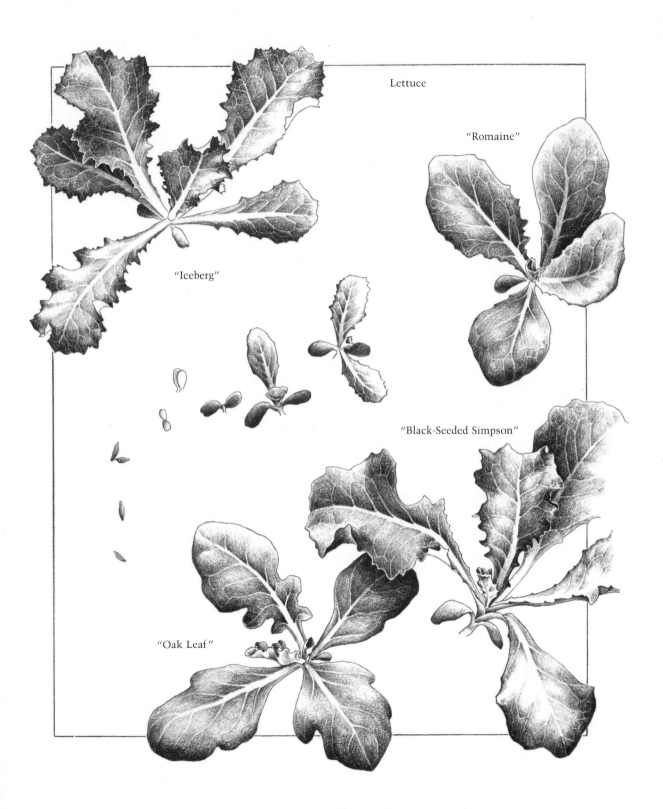

Lettuce

"Romaine"

"Iceberg"

"Black-Seeded Simpson"

"Oak Leaf"

Endive Escarole

Compositae
  Cichorium endivia

Endive and escarole are the two kinds of *Cichorium endivia* commonly grown in North America. These pungent-tasting leafy plants are closely related to chicory, *Cichorium intybus.* Chicory, the same blue-flowered plant that grows wild on our roadsides, is in its cultivated state known as Witloof or French Endive. Its seedlings will at first resemble the endive and escarole.

The culture of endive and escarole probably began somewhere in the eastern Mediterranean or Asia Minor. They were certainly eaten by the Egyptians and Greeks before the Christian era. By the thirteenth century both plants were grown in Europe, and from there were brought to North America in the 1600s.

Endive and escarole join lettuce in the family *Compositae.* As can be expected, they all look quite the same when they make their first appearance with a pair of small, tender, round cotyledons. Even as the first leaves unfold there is much opportunity for confusion, since there are so many lettuce cultivars with similar leaf shapes. Cases of mistaken identity are most apt to occur between the heavily toothed forms of lettuce, like the "Iceberg" varieties, and the fringed and curled leaves of the endive. The broad, round leaves of the escarole soon grow to be much larger than those of any lettuce seedlings. On very close examination one will notice fine hairs on the underside of the leaves of the *Cichoriums,* and these leaves are not quite as tender as those of the lettuce.

Endive and escarole should appear in about ten days from planting.

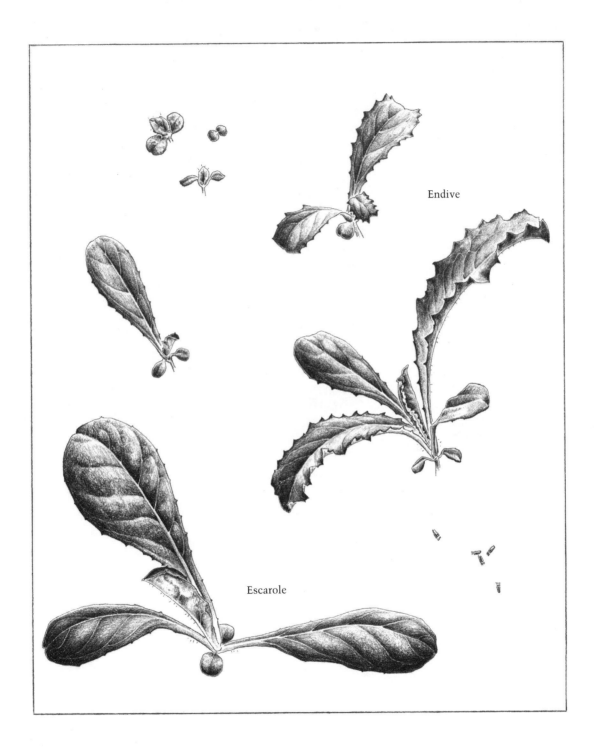

Endive

Escarole

Salsify

Compositae
  Tragopogon porrifolius

Salsify is often called "oyster plant" or "vegetable oyster" because the flavor of its slender, parsnip-like root is similar to that of the shellfish. Salsify was known to the ancient civilizations of the Mediterranean both as a food and as a medicinal plant; it was gathered from the wild then and not deliberately cultivated until about two thousand years ago. It has been grown widely in Europe since about 1600 for its ornamental flowers as well as for its interesting flavor.

Black salsify, more properly called scorzonera, is beginning to appear more often in seed catalogues. This plant is also a member of the family *Compositae,* and the seedlings are very similar. Scorzonera, *Scorzonera hispanica,* is a perennial grown for its edible, dark-skinned roots.

In their early youth, both salsify and scorzonera seedlings are much in danger of being weeded out of the garden with the grasses. Grasses are monocotyledonous, whereas these two vegetables have two "seed leaves." Although this is a major distinction to a botanist, it is not an especially obvious characteristic in the field. The salsify pictured opposite is a slow-growing seedling. All of its slightly folded leaves grow up out of a basal crown; the leaves are hairless, the base slightly reddish. Many grass weeds are slightly fuzzy on the underside of the leaves and most send up lateral shoots or runners.

Salsify seeds germinate poorly and are not long-lived; it is advisable to purchase them fresh every year. Good seed should produce seedlings in seven days.

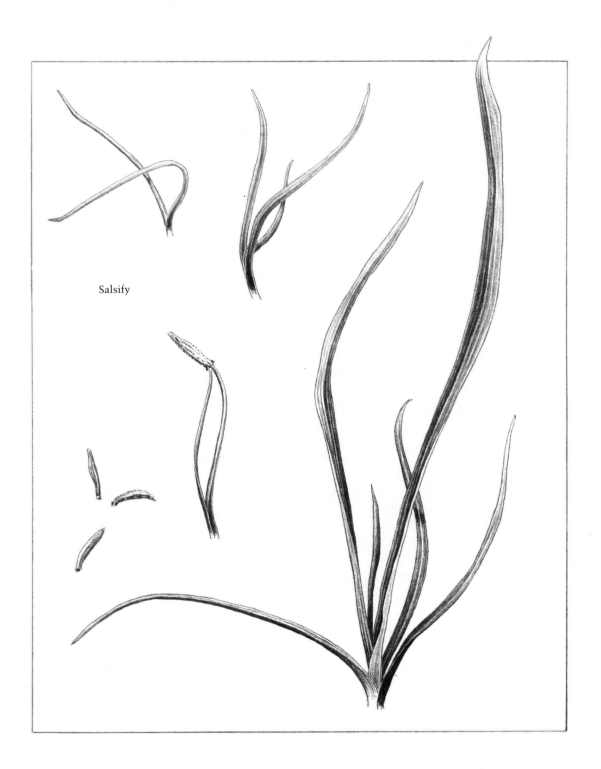

Salsify

Globe Artichoke

Compositae
  Cynara scolymus

The globe artichoke, sometimes called the "green" or "French" artichoke, is another plant that few can recognize as a seedling. Artichokes are grown as perennials and are most often propagated from offshoots of established crowns. The reason for this is that plants grown from seed may not "come true," meaning that the resulting plant may not imitate the parent plant. Most seed companies do offer seed for those who wish to experiment with these temperamental vegetables.

The artichoke originated in the western and central Mediterranean lands. It is not to be confused with the Jerusalem artichoke, which did not come from Jerusalem and cannot properly be called an artichoke. The girasole or Jerusalem artichoke, *Helianthus tuberosus*, is an edible tuber native to North America.

The artichoke is a later adaptation of the ancient plant cardoon, the young stalks and tender leaves of which are relished. The globe artichoke was developed by the Italians in the beginning of the fifteenth century. It became very popular, especially in France and Spain, but has always remained a luxury plant due to its demanding climatic requirements. In this country it is principly grown in Louisiana, which was settled by the French, and in those parts of California that were settled by the Spanish.

The cardoon and the globe artichoke are close relatives of the thistles. This is not immediately apparent in the case of the artichoke seedlings; the cotyledons and the first leaves have a downy gray-green quality that is quite friendly. There are numerous weed plants, for instance the common mullein, of which the young artichoke is reminiscent. By the time a half dozen or so leaves have formed one begins to see a more jagged outline in the leaf, and small hostile spines appear.

Artichoke seedlings usually emerge in twelve days, but they may take a bit longer.

Globe Artichoke

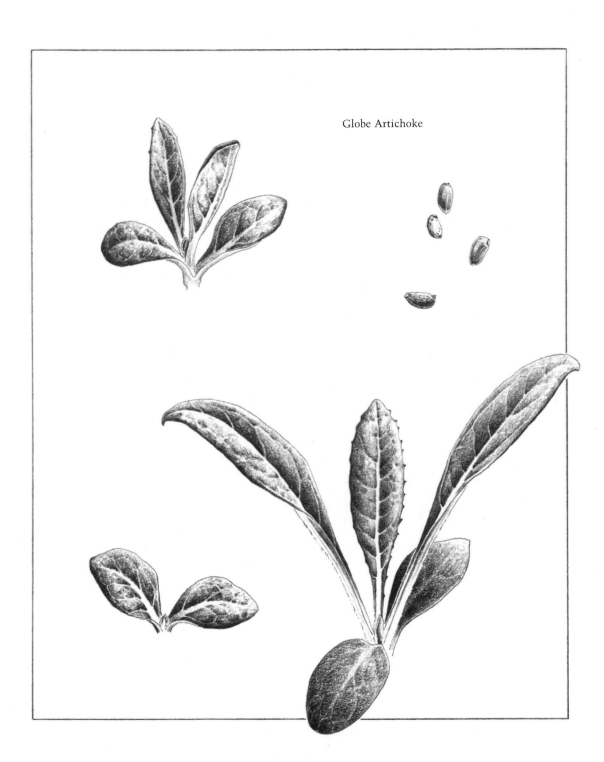

German Camomile

Compositae
  Matricaria chamomilla

Two members of the family *Compositae* are known as camomile. They share so many physical characteristics and uses that there has always been much confusion as to their identification and accurate naming. Illustrated is the Hungarian or German camomile. Some know *Anthemis nobilis*, sometimes further described as English or Roman camomile, as camomile also. To add to the confusion, camomile can be correctly spelled "chamomile."

The apple-like fragrance, daisy-like flowers, and feather-like foliage are common to both plants. Small differences in their flowering parts and habits are what cause them to be placed in different genera. The flowers of the low-growing Roman camomile are popularly used for tea. German camomile grows to be a taller plant, reaching two feet; it has somewhat coarser leaves than the Roman camomile and is less strongly scented. It too is used for tea and is favored for medicinal purposes.

Unfortunately, many seed companies do not specify which type of camomile they are selling. One important clue is that the Roman camomile is described as a perennial whereas the German camomile is an annual. By far the most frequently offered seeds are those of the German camomile depicted opposite. Roman camomile is more often propagated by cuttings, as its seeds cannot be guaranteed to produce plants that have the preferred double flowers.

As might be guessed from its dust-sized seeds, German camomile produces disconcertingly small seedlings. The cotyledons appear as a pair of round leaves. The next leaves are yellow-green narrow-pronged fronds. These seedlings look like none of our other common herbs, but they do bear a strong resemblance to a number of common weeds, most notably those wild camomiles or mayweeds of the same genera, *Anthemis* and *Matricaria*.

German camomile has a poor germination rate which further declines after the first year. It is one of those rare seeds that requires a certain amount of light for germination. Sow these seeds no more than ⅛ inch deep in a light soil. The seedlings should appear in about two weeks.

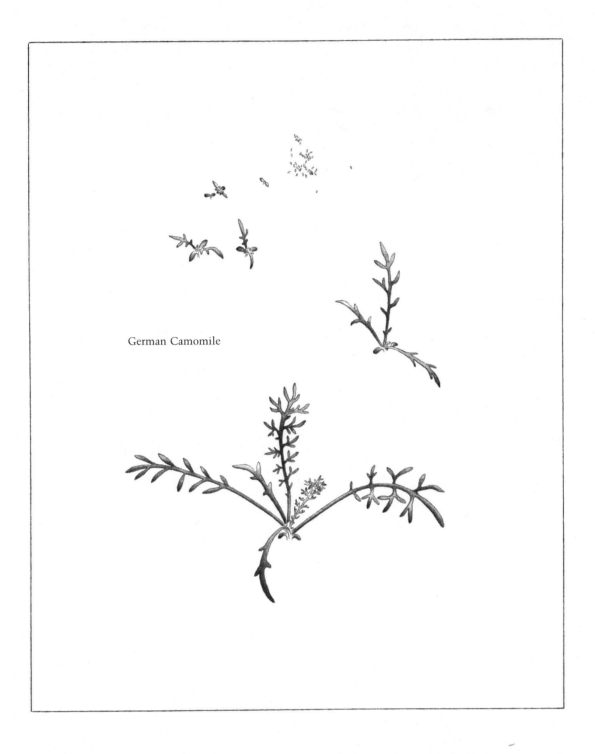

German Camomile

Russian Tarragon

Compositae
  Artemisia dracunculus

The tarragons are believed to have been gathered first in the Himalayas, the Orient, and parts of eastern Europe. Although some cultures found elaborate uses for these plants as early as the first century A.D., Europe seems not to have much favored it until after the twelfth century. It is a recent arrival to the United States, compared to most of our other now-common herbs, having only been offered for sale since the end of the Revolutionary War.

There are two major forms of tarragon, both fairly hardy perennials. French tarragon is propagated almost entirely by cuttings as it rarely produces viable seed. Russian tarragon is the tarragon most frequently offered for sale as seed in catalogues and stores. It is considered by many to be somewhat inferior to the French tarragon. The leaves of both are similar in form but the Russian tarragon is generally a larger plant.

Tarragon emerges as a very tiny pair of round leaves. Next seen are narrow, sharply pointed leaves of a bright yellow-green color. Later leaves occasionally fork but remain quite simple even on the mature, woody-stemmed plant. It is this uncomplicated appearance and small stature in combination with its faint hay-scented fragrance that sometimes cause the tarragon seedling to be mistaken for some insignificant weed.

Tarragon seedlings should appear in about ten days.

Russian Tarragon

Bibliography

Bubel, Nancy. *The Seed Starter's Handbook.* Emmaus, PA: Rodale Press, 1978.

Camp, Wendel H.; Boswell, Victor R.; Magness, John R. *The World in your Garden.* Washington, DC: National Geographic Society, 1957.

Edlin, H. L. *Plants and Man: The Story of our Basic Food.* Garden City, NY: The Natural History Press, 1969.

Halpin, Anne Moyer, ed. *Unusual Vegetables: Something New for this Year's Garden.* Emmaus, PA: Rodale Press, 1978.

Hyams, Edward. *Plants in the Service of Man.* Philadelphia, PA: J. B. Lippincott Co., 1971.

Hylton, William, ed. *The Rodale Herb Book.* Emmaus, PA: Rodale Press, 1974.

Janick, Jules. *Horticultural Science.* San Francisco, CA: W. H. Freeman and Co., 1979.

Lovelock, Yann. *The Vegetable Book: An Unnatural History.* New York, NY: St. Martin's Press Inc., 1972.

Philbrick, Helen; Gregg, Richard B. *Companion Plants and How to Use Them.* Old Greenwich, CT: Devin-Adair Publishing Company, 1966.

Schery, R. W. *Plants for Man.* Englewood Cliffs, NJ: Prentice-Hall Inc., 1972.

Schwanitz, F. *The Origin of Cultivated Plants.* Cambridge, MA: Harvard University Press, 1966.

Smith, A. W. *A Gardener's Book of Plant Names.* New York, NY: Harper and Row Publishers, Inc., 1963.

Thompson, Homer C.; Kelly, William C. *Vegetable Crops.* New York, NY: McGraw-Hill Inc., 1957.

Tyler-Whittle, M. S. *The Plant Hunters.* Radnor, PA: Chilton Book Co., 1970.

Whitaker, T. W.; Bohn, G. W. "The Taxonomy, Genetics, Production and Uses of the Cultivated Species of Cucurbita". *Economic Botany* 4:52-81, 1950.

Notes

Garden Plan

What Was Planted	When	Expected Emergence Date

What Was Planted	When	Expected Emergence Date

JUDITH ELDRIDGE

is an artist with a degree in printmaking from Massachusetts College of Art, and exhibits her work throughout New England. She has taken botany courses at the University of New Hampshire and has gardened since age ten—although she personally would eat nothing then but potatoes and carrots. Eldridge now lives in Boylston, Massachusetts.

CABBAGE OR CAULIFLOWER?

has been set in a film version of Trump Medieval, a typeface designed by Professor Georg Trump in the mid-1950s and cast by the C. E. Weber Type-foundry of Stuttgart, West Germany. The roman letter forms of Trump Medieval are based on classical prototypes, but have been interpreted by Professor Trump in a distinctly modern style. The italic letter forms are more of a sloped roman than a true italic in design, a characteristic shared by many contemporary typefaces. The result is a modern and distinguished type, notable both for its legibility and versatility.

The text and cover were designed by Anne Chalmers and composed by DEKR Corporation, Woburn, Massachusetts. The paper is Warren's #66 Antique, an entirely acid-free sheet. Maple-Vail Book Manufacturing Group, Binghamton, New York, was the printer and binder.